Intangible Materialism

Intangible Materialism

The Body, Scientific Knowledge, and the Power of Language

Ronald Schleifer

University of Minnesota Press Minneapolis / London

Every effort was made to obtain permission to reproduce material in this book. If any proper acknowledgment has not been included here, we encourage copyright holders to notify the publisher.

"Meditation on Certainty," by Jacques Roubaud, from *Some Thing Black*, trans. Rosemarie Waldrop (Elmwood Park, Ill.: Dalkey Archive Press, 1990).

"Complex Motor Tick," by Emma Morgan, from *A Stillness Built of Motion: Living with Tourette's* (Florence, Mass.: Hummingwoman Press, 1995).

"Tortoise Shout," by D. H. Lawrence, from *The Complete Poems*, ed. Vivian de Sola Pinto and Warren Roberts (New York: Viking Press, 1964).

"Ash Wednesday," by T. S. Eliot, from *Selected Poems* (New York: Harvest/HBJ Book, 1964).

"A Deep Sworn Vow," by W. B. Yeats, from *The Collected Poems of W. B. Yeats* (New York: Macmillan, 1956).

"This Living Hand," by John Keats, from *English Romantic Writers*, ed. David Perkins (New York: Harcourt Brace, & World, 1967).

"Connoisseur of Chaos," by Wallace Stevens, from *The Collected Poems of Wallace Stevens* (New York: Alfred A. Knopf, 1954). Copyright 1954 by Wallace Stevens and renewed 1982 by Holly Stevens. Reprinted by permission of Alfred A. Knopf, a division of Random House.

"Episode of Hands," by Hart Crane, from *The Complete Poems and Selected Letters and Prose of Hart Crane*, ed. Brom Weber (New York: Liveright Publishing Corporation, 1952). Copyright 1933, 1958, 1966 by Liveright Publishing Corporation. Copyright 1952 by Brom Weber. Reprinted by permission of Liveright Publishing Corporation.

Excerpts of the Introduction were previously published in *Encyclopedia Americana* 22 (1998): 559–60; reprinted with permission from Scholastic, Inc. Passages of chapter 1 previously appeared in "The Logic of Diagnosis," by Ronald Schleifer and Jerry Vannatta, *Journal of Medicine and Philosophy* 31 (2006): 363–85; reprinted with permission from Oxford University Press. Chapter 3 is a revised version of "The Poetics of Tourette Syndrome: Language, Neurobiology, and Poetry," *New Literary History* 32 (2001): 563–84.

Published by the University of Minnesota Press
111 Third Avenue South, Suite 290
Minneapolis, MN 55401-2520
http://www.upress.umn.edu

Library of Congress Cataloging-in-Publication Data

Schleifer, Ronald.
Intangible materialism : the body, scientific knowledge, and the power of language / Ronald Schleifer.
p. cm.
Includes bibliographical references and index.
ISBN 978-0-8166-4467-4 (hc : alk. paper) — ISBN 978-0-8166-4468-1 (pbk. : alk. paper)
1. Materialism in literature. 2. Materialism. 3. Body, Human, in literature. 4. Body, Human (Philosophy). 5. Literature and science. 6. Literature—Philosophy. 7. Semiotics. I. Title.
PN56.M35S35 2009
809'.9338—dc22

2009006992

Printed in the United States of America on acid-free paper

The University of Minnesota is an equal-opportunity educator and employer.

18 17 16 15 14 13 12 11 10 09 10 9 8 7 6 5 4 3 2 1

For Cyrus and Benjamin

Contents

Acknowledgments

In chapter 2 of this book, I cite Stephen Jay Gould's remarkable description of the encyclopedic procedure of Darwin's work. "Whenever he introduces a major subject," Gould writes, "Darwin fires a volley of disparate facts, all related to the argument at hand—usually the claim that a particular phenomenon originated as a product of history. . . . Darwin's greatest intellectual strength lay in his ability to forge connections and perceive webs of implication (that more conventional thinking in linear order might miss). When Darwin could not cite direct evidence for actual stages in an evolutionary sequence, he relied upon consilience—and sunk enough roots in enough directions to provide adequate support for a single sturdy trunk of explanation." The varied roots Darwin sinks are beyond the ability of most people, and the rest of us have to depend on the patience and wisdom of our friends and colleagues in order to support the branches of explanation we pursue. More than usual, I think, I have benefited from such people in order to gather my argument and organize my thinking.

Chief among my helpers on this project have been my sons, Benjamin and Cyrus, to whom I dedicate this book. Ben, who studies neuroscience, read through many parts of the manuscript and kept me from many (though I fear not all) mistakes about science. His interest in the neurological bases of music—and also, I hope, his joyful sense of experience—can be seen throughout this book. Cy works in the sociology of religion; much of chapter 5 benefited from his thoughtful ecumenical sense of the place and power of religion in people's lives, and the whole has benefited from his generous ability to comprehend and respect what others hold dear. It's a great pleasure to learn from our children—and even greater to share in common pursuits with them. A good deal of the fun and fulfillment of working on *Intangible Materialism* stems from the three of us sharing our different interests concerning many of the issues examined in this book.

Many others shared thoughts about these and other issues. James Hawthorne is chief among my supportive friends on this project: his knowledge of philosophy of mind helped guide early chapters, and his generous personal correspondences, which I have liberally called on throughout this book (even when we didn't agree), and his careful emendations have repeatedly made the organization of my argument clearer. Tania Venediktova helped me devise the current title, which replaced an unspeakable earlier title I mention in chapter 1; Robert Markley's conversation and friendship enriched and continues to enrich my thinking, keeping me precise and also lively; I appreciate Daniel Cottom's general goodwill and precision for details of argument; Russell Reising's imagination and excitement make work fun; and the intelligent good cheer of Nancy West benefits all around her. Still others who contributed to this work include my colleagues at the University of Oklahoma—Susan Kates, Alan Velie, Eve Bannet, Vince Leitch, Tim Murphy, Peter Barker, Ellen Greene, Henry McDonald, Francesca Sawaya, Jonathan Stalling, Christopher Carter, and Lawrence Frank, all of whom often contributed to my sense of the shape of my argument without their knowing. On the other hand, the director of the University of Minnesota Press, Doug Armato, seemed to know better than I did, encouraging this project to completion with remarkably consistent and patient support. *Intangible Materialism* benefited greatly from the readers he chose, Andrzej Warminski and another anonymous reader, whose thoughtful and sympathetic criticism clarified and enlarged my vision. Finally, friends and coauthors Sheila Crow, Robert Con Davis, J. L. Jacobs, Daniele McDowell, Nancy Mergler, Jerry Vannatta, Alan Velie, and Nancy West contributed in substantial ways to this project.

They have done so because much of the working out of the argument of this book is built on—though I hope in no "reductionist" way—and developed from various intellectual pursuits I have undertaken both seemingly by myself and explicitly with others over many years. My interest in semiotics began with the translation (with Daniele McDowell and Alan Velie) of A. J. Greimas's *Structural Semantics* and my subsequent book on Greimassian semiotics, *A. J. Greimas and the Nature of Meaning*. Greimas marks important moments in the present argument. The situation of semiotics as a science is further elaborated in *Culture and Cognition,* which I coauthored with my longtime friend and colleague R. C. Davis and my other longtime friend, my wife, Nancy Mergler. The practical semiotics of a "science"—the practical use of narrative theory—is pursued in *Medicine and Humanistic Understanding,* a DVD-ROM publication that I coauthored with Jerry Vannatta, M.D., and Sheila Crow; in working on this project, we had the pleasure of interviewing doctors and scholars throughout the United States. In *Rhetoric and Death,* I developed a notion of "negative materiality" in the context of the examination of the rhetoric of modernism, and here I pursue a sense of the "negative science" of semiotics that attempts to situate such negative materiality in relation to empirical sciences rather than rhetoric. The rhetorical examination is also pursued in *Criticism and Critique,* another book I coauthored with R. C. Davis. In a parallel manner, *Modernism and Time* attempts, with its focus on the early twentieth century, to historicize the kind of approach to experience embodied in the "scientific knowledge" explored here, much as Mary Poovey (as I note in chapter 1) situates the advent of a postmodern notion of fact in the years around the turn of the twentieth century. In *Modernism and Time,* I do this in relation to what I call "non-transcendental disembodiment" of phenomena such as electricity, finance capital, quantum physics, and literature. In that book, I did not mention another material phenomenon of the late nineteenth century affecting discursive sensibility, popular photography, but I did examine this topic in an essay I wrote with Nancy West, "Photography, Commodities, and Post-Romantic Discourses in Hardy and Stevens," which, like chapter 3 in this book, attempts to situate poetry in a homologous relationship to material reality. Finally, the homological method of this study pursues the analogical thinking I examine in the book by that name, *Analogical Thinking,* though here, as I note in the Introduction, I have learned what Stephen Jay Gould rightly describes as the "material reality" of homologies. I call on many of these previous works of mine—and others as well—with references and a small number of passages adapted from them to make the present argument. I thank the editors and publishers of previous publications for permission to revise and reuse these texts in the present book.

Much of this book has been presented at public forums: meetings of the Society for Literature, Science, and the Arts in Pasadena, Amsterdam, Chicago, Brussels, and Norman, Oklahoma; the History of Science Colloquium Series at the University of Oklahoma; the International Society for the Study of European Ideas in Malta and Helsinki; and speaking engagements at Colorado College, the University of Illinois, Moscow State University, the University of Cologne, and the Semiotics Program, University of Quebec. In these venues, the engagement and responses of often anonymous scholars helped to clarify and shape my work on this book. Several presentations made jointly by Jerry Vannatta and me (at a conference exploring pain, literature, and culture in Athens several years ago; at a conference exploring ethics in Stellenbosch, South Africa; and at a forum for health care workers in Moscow), all of which focused on important aspects of medicine (its imbrication with narrative, its dealing with pain and personhood, its traditions of service, its everyday ethics), helped in the development of chapter 5. In chapter 1 I borrow and revise a few pages from an article examining the Peircean logic of diagnosis I coauthored with Jerry. Working closely with Jerry and Sheila Crow on the ways humanistic understanding can inflect scientific practice has been a vital and fulfilling aspect of my career in both teaching and scholarship. In fact, this project began with an examination of Tourette syndrome (now in chapter 3), which grew out of the course on literature and medicine that Jerry envisioned and that we have taught together for many years to undergraduates and medical students. Meetings and interviews for *Medicine and Humanistic Understanding* with Dr. Rita Charon, Dr. Rafael Campo, Dr. John Stone, Dr. Abraham Verghese, and Dr. Oliver Sacks (among others) contributed to my understanding of materialism, science, and literature.

Intangible Materialism, with its focus on faces, voices, hands, and pain, pursues its argument in the contexts of basic facts, evolution that always takes place within communities, and the semiotic promise, as Peirce says, of a future. This book is part of a longer argument I have followed for years with friends, colleagues, and now my sons.

Introduction: Materialist Literature

In his critique of what he calls "the modernist project" and I call Enlightenment modernism, Bruno Latour argues that "matter is not given but a recent historical creation" (1999, 207). "Philosophers and sociologists of techniques," he writes,

> tend to imagine that there is no difficulty in defining material entities because they are objective, unproblematically composed of forces, elements, atoms. Only the social, the human realm, is difficult to interpret, we often think, because it is complexly historical and, as they say, "symbolic." But whenever we talk of matter we are really considering . . . a *package* of former crossovers between social and natural elements so that what we take to be primitive and pure terms are belated and mixed ones. Already we have seen that matter varies greatly from layer to layer—matter in the layer I have called "political ecology" differs from that in the layers called "technology" and "networks of power." Far from being primitive, immutable, and ahistorical, matter too has a complex genealogy. (204–6)

The variance of matter from layer to layer, or among "levels" of understanding, without recourse to one kind of supernaturalism or another is the argument of this book. *Intangible Materialism* attempts to articulate a global sense of materialism in relation to the more or less unproblematic systematics of mechanical analysis, the more or less satisfying historical explanations of natural selection, and the complications of semiotic understandings of what Latour calls the symbolic. To this end, it takes up Charles Sanders Peirce's analysis of meaning and experience in relation to his categories of icon, index, and symbol and A. J. Greimas's hope that the *experience* of meaning in our lives can be understood with the same precision that mathematical physics has taught us to understand the mysteries of the world. "The 'piece of wax' of Descartes," Greimas writes, "is no less mysterious than the symbol of the moon. It is simply that chemistry has succeeded in giving an account of its elementary composition. It is toward an analysis of the same type that structural semantics must proceed" (1983, 65).

Greimas's hope, in the term of Edward Said, is altogether *worldly*, which is to say it is altogether secular in the way that Enlightenment modernism is—after the terrible religious wars of the seventeenth century—in its striving to create a secular order for experience. Said describes such an order in "The Politics of Knowledge" when he writes that "there must be, it seems to me, a theoretical presumption that in matters having to do with human history and society any rigid theoretical ideal, any simple additive or mechanical notion of what is or is not factual, must yield to the central factor of human work, the actual participation of people in the making of human life" (1994, 147). He states further that "this kind of human work, which is intellectual work, is worldly, that it is situated in the world, and about that world" (147) and that it is engaged with culture in an "unprovincial, interested manner" (151). The combination of unprovincial and interested, as I argue in *Analogical Thinking*, conditioned the very experience of modernity, meaning by *modernity* the secular–scientific experience in the West since its beginning in the Enlightenment (2000a, 91). Such secular–scientific experience, as David Chalmers (among many others) has noted, seems to destroy "any sense of fundamental mystery" in the world (1996, 42), and Chalmers's philosophical study of consciousness and experience attempts, as I do here, to *notice* and acknowledge the mysterious in our experience without abandoning worldliness.

The Work of Literature

The process of noticing, sometimes called observing, recording, recognizing, or paying attention, is a subtheme of this book: it is what seems to me the *mysterious* aspect of post-Einsteinian physics in the twentieth century. It requires the abandonment of "any simple additive or mechanical notion of what is or is not factual" (Said 1994, 147); or rather, it requires the reformulation, as Mary Poovey notes in an argument I follow in chapter 1, of the notion of "fact" that "occurred at different moments in different disciplines" near the turn of the twentieth century (1998, 3). Such noticing has always been the work of literature—its *worldly* work, as Said suggests, equipment for living (see Burke 1994). Near the turn of the twentieth century, Joseph Conrad contended that literature aims "by the power of the written word to make you hear, to make you feel—[and] . . . before all, to make you *see*" (1924, xiv). In a way, the phenomenon of such seeing (noticing, or "witnessing") can be taken as mysterious—or at least seemingly charged with what feels like *immediate* significance, the enchantment that George Levine depicts in his engaged study of Darwin and "the re-enchantment of the world" (2006)—without abandoning material worldliness. In fact, literature might properly be defined in relation to the different levels of understanding that Latour mentions and I pursue in this book—the mechanics of technology, the natural history of political ecology, and the semiotics of networks of power—and as such be understood as a *material* phenomenon. The materiality of such an understanding would be alternatively systematic, historical, and symbolic.[1]

Literature, as privileged forms of verbal discourse and patterned uses of linguistic expression that go beyond the pragmatic function of communication, has been a part of all organized human societies. Ancient Chinese culture, for instance, developed the term *wen,* which referred to "patterned" or rhymed language, what we might call literary language or poetry. *Wen* also referred to patterns or markings on natural objects, to inherited cultural traditions, and to the order of the cosmos, and as such it inhabits the "levels" of physical fact, biological adaptations, and contrary-to-fact semiotic assertions examined in chapter 2 and throughout this book. Similarly, classical Greek culture coined the term *poetry* (from the Greek *poesis,* meaning "to make") and, in Plato and Aristotle, developed a sophisticated sense of the function and workings of what would be called literature today. But even earlier preclassical Greek commentaries on poetry and poets, mostly handed down in fragments, suggest that poets' discourse was a privileged mode of linguistic activity. These commentaries describe the power of such discourse to arouse basic emotions of joy, anxiety, or anger; its special role of preserving personal glory and social values; and its divine origin. In ancient India, as we know from commentaries that predate Plato and Aristotle by more than a thousand years, the practice and appreciation of patterned discourse were deeply integrated into the activities of daily life. Thus, the *Ayurveda,* the Indian science of medicine, believed that a perfectly structured couplet could clean the air and heal the sick. Near the end of this book, I point out that Aristotle also saw a relationship between literature and the health of both body and body politic. But even in less expansive societies than ancient China, Greece, or India, in smaller tribes and villages and in societies without writing, the patterned discourses of myth and poetry are ubiquitous within social formations. Native American cultures, for example, consistently distinguish between sacred and secular narratives, and different tribes carefully define the subject matter and the manner of telling of sacred tales.

Literature, like the "matter" Latour describes, lends itself to various definitions. The broadest and perhaps most literal definition is simply anything that appears in print in the sense of the way scholars (in the sciences especially), for instance, describe the literature in their field. Besides this generalizing use, literature has been defined as imaginative, writing in the sense of its not being literally true, what I describe in *Intangible Materialism* as counterfactual or "contrary-to-fact" discourse that grows out of the very place of human communities in the natural order. A second, related definition arose in the late nineteenth century, when the institutionalized study of modern (as opposed to classical and biblical) languages and literatures arose within the newly established comprehensive universities in Europe and America and writers such as Matthew Arnold emphasized the universalizing, "disinterested," and resolutely nonpolitical nature of literature and culture. This conception comes close to the semiotic idea of "shifting out" of discourse I examine in chapter 5, or what the evolutionary psychologist Pascal Boyer (2001) calls, in his examination of the evolutionary origins of religion, "decoupled cognition." An extension of this idea of the disinterested nature of literature emphasizes its nonpragmatic use of language: as W. H. Auden says in his elegy for Yeats, "poetry makes nothing happen" (1966, 142). In some ways, these definitions of literature make it seem intangible.

Additional definitions of literature are particularly connected to activities in the world and to my discussions of the languages of cursing, interpersonal community, and the sacred—discussions that touch upon the power of poetry to arouse basic emotions of joy, anxiety, or anger; its special role of preserving personal glory and social values; and its divine origin. Perhaps the oldest definition of privileged discourse, related to but distinguishable from the definition of literature as imaginary and contrary to fact, sees it as the linguistic embodiment of the sacred and the mysterious within social formations. Walter Benjamin, for instance, describes the power of language to enact revelation within the "conflict . . . between what is expressed and expressible and what is inexpressible and unexpressed" in order to achieve "communion with God's word" (1996, 66, 69). "Language," he says, "is in every case not only communication of the communicable but also, at the same time, a symbol of the noncommunicable" (74). Even those who do not hold beliefs in transcendental powers may understand literature as embodying the immediate emotional or intellectual power that Benjamin describes.

Another definition asserts that literature is a particular kind of writing, distinct from other uses of language. Recently, Michael Trimble (2007) offered a material, neurological analysis of such "other uses" of language closely related to the materiality of religious experience examined in chapter 5. Others throughout the twentieth century pursued a "scientific" linguistic definition—often identified with structuralist analyses, such as those of Roman Jakobson and Greimas—by emphasizing the ways that literature transforms ordinary language by *calling attention* (in its work of noticing) to its linguistic and aesthetic, as opposed to communicative, attributes. In this linguistic conception, one can always find the literary element within any discourse, such as the elements of "literariness" or "poeticity" that the Russian formalists and Jakobson described in the early twentieth century, or such as the "narrative grammars" that Greimas described. The noncommunicable aspect of language—its "music," "poeticity," or, in the language of Peirce, its "iconic" element—calls attention to its materiality.

Closely related to the definition of literature as sacred texts is the assertion that literary texts embody forms of discourse that imitate and modify traditional discourses, especially religious discourses, that a culture has handed down from generation to generation in the same way, for instance, that our earliest ancestors passed on for more than two million years the technology of the hand axe (see Diamond 1992, ch. 2; Mithen 2006, 164). Thus, one might no longer believe that poetry, as the ancient Greeks believed, was literally inspired by the gods or that it is, as Benjamin asserted, a form of sacred naming, but one could still define literature as embodying the social or communal power of those received discursive forms. (This is a definition by means of homology, the version of historical materialism Stephen Jay Gould sees in natural selection, discussed later in the Introduction.) Literature by this definition, then, is fine writing that takes up the forms of more or less sacred writing—epic, parable, the "inspired" utterances of lyric (including traditional curse poetry, which is related to my discussion of Tourette syndrome in chapter 3)—for new and different ends. In this sense, Alastair Fowler, following the great generic descriptions of literature by Cicero, Horace, Sidney, Dryden, and others, calls literature "an order of works" (1982), the kind of network that Latour wants us to notice. Related to this idea is a sense of literature that Stephen Greenblatt describes as language functioning communicatively *beyond* its historical occasions: "Works of art," he writes, "by contrast [with ordinary texts 'most of which are virtually incomprehensible when they are removed from their immediate surroundings,'] contain directly or by implication much of [their cultural] situation within themselves, and it is this sustained absorption that enables many literary works to survive the collapse of the conditions that led to their production" (1995, 227). Rather than emphasizing the noncommunicative element of discourse, literature here is conceived as language offering an "extracommunicative" element. Even E. O. Wilson, in his materialist program of what Levine calls "strong reductionism" (2006, 125), which I examine in chapter 2, notices "three strokes of luck in the evolutionary area . . . seemingly developed beyond their survival needs" (1998, 48).

A final definition of literature describes it as a normative and value-laden term that designates forms of writing that embody the values of the dominant group or class at any particular time. In this definition, literature is a term of normal or preferred values whose function is to support particular interests and ideologies within a society. Thus, for instance, the designation of literature to describe the mostly New England male writers of pre–Civil War America—what F. O. Matthiessen defined sixty years ago as the American Renaissance (1941)—while excluding the extraordinarily popular and influential works of Harriet Beecher Stowe and other women writers, posits particular styles and subject matters as literary while marking others as subliterary. Critics such as Terry Eagleton and Raymond Williams describe literature in these terms as embodying historically local *ideologies* insofar as ideology is not simply a set of conscious beliefs about the world and society but what Williams calls "structures of feeling" (1961), the "modes of feeling, valuing, perceiving and believing" that Eagleton describes as having "some kind of relation to the maintenance and reproduction of social power" (1983, 15). In this sense, literature is a privileged site for studying material history.

This catalog of definitions offers descriptions of literature that emphasize its universal and philosophical focus (Aristotle), its disinterested and expressive nature (Arnold/Greenblatt), its scientific–linguistic definition (Jakobson), its metaphysical power (Benjamin), its conventionality in terms of genre (Fowler), and its historical and ideological functioning (Williams/Eagleton). In addition to these systematic and historical definitions—or emerging out of these definitions—is a sense of literature that can be situated as part of the *material orders* of our physical bodies, our interpersonal and social organizations growing out of the material–biological world, and the epiphenomenon of these orders: our ability to imagine agency in experience. In other words, literature can also be defined in relation to the concept of *materialism* as it functions within scientific discourses and the modes of understanding that I pursue in *Intangible Materialism.*

This book itself grows out of my recent studies of the relationships between literature and medicine and my long-term attempts to create an account of language and discourse that avoids the opposition between matter and spirit. As part of this larger project, *Intangible Materialism* attempts to articulate a way of understanding human experience—including the special and in many ways defining case of literary experience—that might take its place along with the mechanistic–reductionist understandings of mathematical and mechanic physics and the goal-oriented understandings of biology and other environmental sciences. What is key to virtually all of the definitions of literature I have cataloged is a sense that literature, among its activities of attention, takes up and emphasizes Benjamin's noncommunicative aspects of discourse, which makes the material *nonsense* of language—its sounds, its communal social work, its soundings of feelings beyond sense—something that might be understood in terms of a "negative science" as opposed to the "positive sciences" of Enlightenment modernism. The aim of *Intangible Materialism* is to situate all the activities of literary attention—its sensuousness, its constant recognition of our situation in the world, and its creation and instantiation of laws of meaning for cognition and experience—within a materialist understanding of the world and of experience and to provide senses of the ways that the activities of literature can clarify and perhaps expand what we mean by materialism.

Wholeness, Homology, Analogy

For all this, though, *Intangible Materialism* is hardly a "literary" book. It examines the notion and procedures of scientific reductionism, including an examination of what has been meant by the notion of "fact" since the advent of a more or less secular–scientific worldview in the seventeenth century (chapter 1). It explores levels of scientific explanation within the broad category of scientific knowledge, suggesting that such knowledge can be best comprehended in relation to the categories of mechanical–physics analysis, biological–evolutionary explanation, and the complications of semiotics (chapter 2), what John Casti calls physics, semiphysics, and metaphysics (1995, 77–79). After these framing discussions, *Intangible Materialism* examines possibilities of materialist explanation, focusing on crucial aspects of our human bodily existence: faces (chapter 2), voices (chapter 3), hands (chapter 4), and pain (chapter 5). These chapters touch on literature, but as I said, they are hardly literary studies. Instead, they try to use the semiotics of literary discourse to both reflect and inflect the discussions of a widened comprehension of materialism—what I call material formalism—while they pursue an understanding of the material power of discourse. More specifically, chapter 3 examines the material (homological) resources of literariness and poetry as they are revealed in the strange condition of Tourette syndrome, situated as it is between a clearly physiological ailment created by unbalanced neurotransmitters and an equally clear social dysfunction that responds to and takes up particular cultural mores in its symptoms. Chapter 4 examines the material situation of poetry in relation to the roles of gesture and discourse as an adaptive feature for the individual and group in the natural and social world. In this chapter, I pursue the materialism of adaptation in the physical, evolutionary, and semiotic functioning of the human hand. Handwork, especially in contrast to Tourette syndrome or even to pain, creates sociality, and chapter 4 examines the ways that the *interactive* activities of the hand suggest the interactivity of narrative discourse. Finally, chapter 5 examines the overwhelming material fact of pain and its situation—like Tourette syndrome—between physiology and culture. But if Tourette syndrome helps us to see the physiological resources poetry calls upon, the more general discussion of pain in chapter 5 helps us to see the ways that such physiology contributes to the phenomenology of selfhood and subjectivity in the extreme example of the pursuit of pain in religious experience. Thus, chapter 5 explores physiological analyses of pain, evolutionary explanations of religion, and semiotic descriptions of the ways religion takes up the seemingly negative—objectless and adversive—phenomena of pain for discovering meaning and purport in experience.

These chapters are organized in relation to Charles Sanders Peirce's inaugurating insight into the elements of communication and understanding—the elements of semiosis—his catalog of icon, index, and symbol (and of Firstness, Secondness, and Thirdness) as the constituent modalities of experience, knowledge, and language. In chapter 2, I offer what might be called a "Peircean" definition of matter itself. I argue that the best definition of matter and materialism might well refer to phenomena that are organized (in a kind of material formalism) in relation to iconic sensation, indexical resistance, and symbolic comprehension, alternatively ordered in vocabularies of matter, energy, and information or semiotics. Chapter 1 sets forth Peirce's description of the elements of semiosis under the category of modes of being, and throughout the discussions of materialism, I return time and again to his basic categories. Yet *Intangible Materialism* should not be taken to be a Peircean reading of either materialism or literature. Rather, it takes up Peirce's categories as a kind of homology to (or at least isomorphic with) categories suggested by Erwin Schrödinger in his analysis of the relationship between physics and biology, discussed in chapter 2. In fact, *Intangible Materialism* follows the procedure of homological or analogical thinking throughout its work.

Such a procedure might be, as Gould argues, the alternative to the reductive analyses of mechanistic physics that I discuss in chapters 1 and 2.[2] It might, in fact, offer the "method" of discovering a materialism that is *intangible*. "Analogical expressions," David Burrell writes,

> come into play precisely at those points where one wishes to speak of language itself or of the relation between language and the world, and yet realizes that one must have recourse to a language. At these points we need expressions that function within our language but whose serviceability is not restricted to their role within the language. They must function within the language so that we can get our bearings in using them; but they cannot be restricted to that intrasystematic function if we want to be able to use them in speaking of what we can do with the language as a whole. (1973, 224)

Analogy, then—like Peirce's index, examined in chapter 4—traffics between the tangible (touchable) world and the intangibility of meanings. That intangibilities of meaning can have material effects in the world is, as Shoshana Felman writes, "the scandal of the speaking body" (2002), the functional materialism of literature, so to speak, that I include in my description of the work of literature. Such material effects are clear in what Howard Kushner calls "the cursing brain" manifesting itself in Tourette syndrome, and they are clear (though maybe less so) in the shared handiwork of toolmaking, which conceives tools outside the material situation of their necessity and is not "dependent," as Merlin Donald notes, "on immediate environmental reinforcers or contingencies" (1991, 179). Finally, such material effects can be seen in the phenomenon of pain within religious life—what Ariel Glucklich calls "sacred pain."

The analogy Burrell describes is between language and the world and the even more abstract phenomenon of language itself, language as a whole, as if wholeness were both a felt material fact and something that is not quite tangible. Wholeness, as I argue throughout *Intangible Materialism,* is a quality of discourse and might itself be a function of discourse—though it might more simply be a function of "basic awareness" that Donald describes (2001, 183)—and the levels of understanding I discuss in chapter 2 are also methods of apprehending wholenesses. In his remarkable study of the evolution of music, *The Singing Neanderthals,* Steve Mithen cites the linguist Alison Wray, who argues that "the precursor to language was a communication system composed of 'messages' rather than words" (Mithen 2006, 3), which is to say wholes that do not lend themselves to analysis into constituent parts. In this way, such a communicative system consisted of "vocal gestures," as Mithen calls them—similar to the protolanguage of gestures Donald suggests (1991) and, in fact, to Peirce's index.

Jared Diamond describes the relation of wholeness to natural selection as well as to language and semiosis, focusing on the question—which E. O. Wilson also asks, as we shall see in chapter 2—of "the mystery of what sets the maximum limit on any advantageous trait." Since people "who are big or smart or can run fast" clearly have an advantage, he notes, "why didn't we evolve to become on the average even bigger, smarter, and faster than we now are?" He answers that "natural selection acts on whole individuals, not on single parts of an individual" (1992, 127; for Gould's analysis of "individuals" in the evolutionary theory, see chapter 4, note 22). And Mary Poovey, in her powerful study of *A History of the Modern Fact,* suggests ways that wholeness might inflect our sense—and the reality—of brute facts, "particulars" that seem not to be "whole" but simply given. The point of discovering wholeness on the levels of language, natural selection, and "physical" fact—the levels explored in chapter 2—is to find ways of imagining phenomenology to be part of all these things, to be, despite its seeming experiential subjectivity—or precisely *in* its seeming experiential subjectivity—a material phenomenon itself.[3] Phenomena apprehended as wholes—whether they are facts in the world or ideas about the world—are the stuff of analogy, and the work of this book is to conceive of such phenomena, and the phenomenology of experience altogether, as worldly and material.

Perhaps a better term than analogy—more worldly, more material—is homology, since the term itself layers areas of application: the *Oxford English Dictionary* defines it as a term of art in mathematics, chemistry, biology, and medicine, all fields touched on in *Intangible Materialism.* The biological definition is perhaps most useful:

> 2. a. *Biol.* Having the same relation to an original or fundamental type; corresponding in type of structure (but not necessarily in function); said of parts or organs in different animals or plants, or of different parts or organs in the same animal or plant. (Distinguished from *analogous* . . .)

Under the term *analogy,* the OED notes that "parts which correspond in their real nature (their origin and development) are termed 'homologous'; those which agree merely in appearance or office are said to be 'analogous.'" Homology, then, pursues the relation between language and the world, the "real nature" of things, while analogy pursues appearance. This difference is clear to Gould, whose examination of the "homological reasoning" of narrative I cite in chapter 4. "Similarities may be homologies, shared by simple reason of descent and history," he writes, "or analogies, actively developed (independently but to similar form and effect) as evolutionary responses to common situations. . . . Homologous similarity is the product of history; analogy, as independent turning to current circumstance, obscures the path of history" (1986, 66; see also 2002, 601).[4]

Homologous correspondences in structure but not function govern the relationships among the levels of understanding examined in chapter 2, and such correspondences govern the relationships among the physical, biological, and semiotic understandings of faces, voice, hands, and pain pursued throughout this book. But in another sense, the distinction between homology and analogy so clear for Gould in Darwin's "historical science" is obscured in semiotics where, in the words of Greimas, "Units of communication with different dimensions can be at the same time recognized as equivalent" (1983, 82). Such "recognition" is a result of uses of language *beyond* its structure; it is a result of the future-oriented purposes (or purport) of language that, contrary to fact, seemingly go beyond the material reality of homology that Gould examines in Darwin (1986, 68). Burrell's double description of language as a self-contained system whose function is to interact with the world, like Greimas's description of language, also goes beyond—while encompassing—Gould's materiality reality. In a way, the double definition in both Greimas and Burrell can be seen in the juxtaposition of Ferdinand de Saussure and Peirce in chapter 1, but more significant, I think, is Casti's observation, emphasizing wholeness, that "meaning is bound up with the whole process of communication and doesn't reside in just one or another aspect of it. As a result, the complexity of a political structure, a national economy or an immune system"—he could have added here, semiotics systems more generally—"cannot be regarded as simply a property of that system taken in isolation. Rather, whatever complexity such systems have is a joint property of the system *and* its interaction with another system" (1995, 269). It is just such interactions—the interaction I describe in chapter 1 of different *kinds* of material facts—that allow, I believe, a conception of materialism without reduction.

Intangible Materialism is a book of homologies: hands, for instance, remain constant in structure even when their functions vary across levels of understanding, and their very functionality is both real and, on some level, intangible. Toward the end of this book, I cite Ariel Glucklich describing the nature of biological systems. "From a biological point of view," he writes,

> the human organism consists of numerous systems and subsystems of cellular organization. At the most basic level are single cells, then individual organs, and moving up the scale there are complex functional systems such as the circulatory system, the reproductive system, the immune system, the nervous system, and so forth. In calling

> a unit a "system" we are making several assumptions. For instance, the separation into systems is functional rather than anatomical. The same organs can be "shared" by more than one system. The organizational principle implies at least three critical features: communication within the system, the notion of telos or systemic goal, and isomorphism between different levels of subsystems. (2001, 91)

Glucklich defines isomorphism, following Gestalt theory, as "similarities" that can be *seen* but that can also be understood as parallel responses—which are *noted* or *judged* rather than seen—to the physical–material environment, and they can also be understood as phenomena that are *taken up* rather than seen in relation to the ends of communication. In fact, the critical features he isolates in defining systems—similarities apprehended in relation to physical–material reality, apprehended in relation to goals, and apprehended in the context of possibilities of communication—correspond to the three levels of understanding examined in this book: physics, biology, semiotics.

What semiotics teaches us—and what the elaborate and perhaps defining example of semiotics, literature, also teaches—is that phenomena can be fully conceived within a physical–material environment and still possess the effect of intangibility. This is most clear, I think, in the ways evolutionary psychology analyzes religion, and especially supernatural agency, as discussed in chapter 5. But it can also be seen in tools and the contrary-to-fact narratives that allow tools to be created and in articulations of impassioned speech treated in the other chapters. Homologies (and analogies) do not lend themselves to the reductive explanations I examine in chapter 1, nor do they situate themselves outside the material environment. Instead, as I argue, they widen the sense of materialism to include phenomena that are real and worldly without being tangible. Literature can help us to understand this intangible materialism, just as the examination of scientific knowledge and scientific "facts" can help us understand literature. This complex understanding, in any case, is my hope for this book.

1. Intangible Materialism

> Information is information, not matter or energy. No materialism which does not admit this can survive at the present day.
>
> —*Norbert Wiener,* Cybernetics

> Matter is not given but a recent historical creation.
>
> —*Bruno Latour,* Pandora's Hope

Intangible Materialism examines the possibility of a conception of materialism and matter beyond the reductionism of Cartesian mechanics and its heirs. It aims at substantiating Norbert Wiener's assertion that information is as basic to any contemporary concept of materialism as matter and energy are to that concept in classical science. Wiener's global aim is to account for a complex materialist worldview without recourse to notions of "ghosts" in the machines of matter and without recourse to an absolute mechanistic reductionism that seems to blind itself to a host of *material* phenomena. To this end, *Intangible Materialism* focuses on semiotics, information theory, and the framework of Charles Sanders Peirce's conceptions of the modalities of understanding in relation to physics and physiology, biology and evolutionary science, and semiotics and literature. The chapters of this book are organized around particular material aspects of human life: faces, voices, hands, and the phenomena of pain. Each chapter draws links between material sciences—mechanistic physics, neurobiology, evolutionary biology, anthropology—and literary semiotics.

In a way, the aim of these chapters (and the book as a whole) is to situate the *activities* of literature—reading, listening, or communicative gesturing (noticing) and the cognitive and phenomenal responses to these activities—within a materialist understanding of the world and of experience. It also provides, I hope, senses of the ways that the activities of literature can clarify and perhaps expand what we mean by materialism altogether.

Reductionism

Originally, I drafted this book with the ungainly title of *Materialism without Reduction,* and in this title I was thinking of the controversies, in science and philosophy, over whether or not any working notion of materialism *requires* a sense that all phenomena must be reducible to a physical or "physicalist" conception of matter. As David Chalmers argues in *The Conscious Mind,* "the widely held doctrine of *materialism* (or *physicalism*) . . . is generally taken to hold that everything in the world is physical, or that there is nothing over and above the physical, or that the physical facts in a certain sense exhaust all the facts about the world" (1996, 41). Chalmers describes what he calls the nature of reductive explanation. "The remarkable progress of science over the last few centuries," he writes,

> has given us good reason to believe that there is very little that is utterly mysterious about the world. For almost every natural phenomenon above the level of microscopic physics, there seems in principle to exist a *reductive explanation:* that is, an explanation wholly in terms of simpler entities. In these cases, when we give an appropriate account of lower-level processes, an explanation of the higher-level phenomenon falls out.
>
> Biological phenomena provide a clear illustration. Reproduction can be explained by giving an account of the genetic and cellular mechanisms that allow organisms to produce other organisms. Adaptation can be explained by giving an account of the mechanisms that lead to appropriate changes in external function in response to environmental stimulation. Life itself is explained by explaining the various mechanisms that bring about reproduction, adaptation, and the like. Once we have told the lower-level story in enough detail, any sense of fundamental mystery goes away: the phenomena that needed to be explained have been explained. (42)

Chalmers further argues that most natural phenomena are susceptible to such explanation: he mentions the explanation of heat in physics, the phases of the moon in astronomy, earthquakes in geophysics, and even "a phenomenon such as learning" in cognitive sciences, which can be explained, as he says, in terms of "various functional mechanisms—the mechanisms that give rise to appropriate changes in behavior in response to environmental stimulation" (42).[1]

Chalmers argues that such "materialist" explanations do not adequately account for the phenomenon of consciousness, the *felt* experience of the world that he often describes in terms of "meaning."[2] For Chalmers, *consciousness*—including, by implication, the *experience* of meaning—is immediate and unanalyzable: it is not subject to scientific (e.g., *semiotic*) analysis but is simply a given, a "further fact" beyond any accompanying physical processes (107). That is, he seems to have the intuition—he claims throughout that conscious experience itself is necessarily "intuitive"—that the felt qualities of experience such as the color red, harmonies of sound, or pain (146) are not reducible to the mechanical explanations of materialism; "There is no special reason," he writes, "to believe that there will be a single isolable mechanism that underlies experience" (241). Thus, he notes that "all it means to be a conscious experience, in any possible world, is to have a certain feel," and he adds that Saul Kripke "makes a similar point, although he puts the point in terms of essential properties rather than in terms of meaning" (133). "Whether or not consciousness *is* a biochemical structure," he argues earlier, "that is not what 'consciousness' *means*" (106):

> And almost all the high-level phenomena that we need to explain ultimately come down to structure or function: think of the explanation of waterfalls, planets, digestion, reproduction, language. But the explanation of consciousness is not just a matter of explaining structure and function. Once we have explained all the physical structure in the vicinity of the brain, and we have explained how all the various brain functions are performed, there is a further sort of explanandum: consciousness itself. Why should all this structure and function give rise to experience? The story about the physical processes does not say. . . .
>
> The fact that consciousness accompanies a given physical process is a *further fact,* not explainable simply by telling the story about the physical facts. In a sense, the accompaniment must be taken as

> brute. . . . There will always remain an element here that is logically independent of the physical story. (107)

The inclusion of language on his list of high-level phenomena is striking, because it assumes that the experience of language does not give rise to what I describe as "felt meaning," a kind of seeming *intuition*. Or perhaps Chalmers has in mind here a definition of language as cognitive process functionally implemented in the mechanism of the brain, as distinct from "the felt experience of linguistic meaning" I will describe. Given the phenomenon of "consciousness itself" that is "logically independent of the physical story," Chalmers argues against a universal materialism and in favor of a "naturalistic dualism." He calls the dualism of physical fact and experiential phenomenology "naturalistic" because he rejects, as I do, any "supernatural" account of experience, though he argues, unlike the argument that follows in this book, that materialism is the wrong term to use for phenomenology.[3]

Chalmers's description of experience as wholly—and unselfconsciously—understood as *individual* experience betrays a kind of parochialism. In contrast to Chalmers's assumption, Mary Poovey offers a history of the "reconceptualizing of 'experience'" in the seventeenth century. "As Peter Dear has argued," she writes, "the Aristotelian concept of experience emphasized both commonplaces and the communities in which what constituted a commonplace was adjudicated. 'An "experience" in the Aristotelian sense was a statement about *how things happen* in nature rather than a statement of *how something had happened on a particular occasion*,' Dear writes. 'For Aristotle, the nature of experience depended on its embeddedness in the community; the world was construed through communal eyes.' By the early seventeenth century, a new concept was beginning to rival the Aristotelian truisms. This new way of understanding experience stressed the particularity of individual events and individual observers" (1998, 70–71; she cites Dear 1995, 4, 23). In this, Poovey emphasizes a conception of experience as mediated rather than immediately intuitive.[4]

The issue for both Chalmers and Poovey, I think, is how to account for the phenomenology of experience. This has been the problem for poetics and literary studies throughout our tradition in the West. In the last century in particular, semiotics has attempted to examine and analyze the phenomenology of meaning, as A. J. Greimas has done, in terms of the *cognitive* sense of what a phrase or sentence might mean—which we seemingly apprehend as directly (intuitively?) as a color or a taste[5]—and also the "felt sense" of confusion or bewilderment, of elation or simple contentment, the *phenomenal* experience of the failure or the success of grasping a meaning. Thus, in *A. J. Greimas and the Nature of Meaning,* I argue that for Greimas, "the 'nature' of meaning is *phenomenal;* it 'exists' as the felt sense of its presence, a signifying *whole* beyond the limits of the sentence, or the felt sense of its negated presence, the 'nonsense' and 'bewilderment' of fragmented sense" (Schleifer 1987, xix). Moreover, as I note in this book, Peirce attempts to situate and comprehend, perhaps more basically, the phenomenology of color experience—like Chalmers, he uses red as his example—in terms of his semiotic category of the "icon."

This semiotic analysis of meaning suggests—as does the working "semiotics" of reading and responding to literary texts—that a philosopher like Chalmers too quickly takes for granted that "meaning" and the phenomenal qualities of "meaningful" experience are immediate, intuitive, simply given. It is the presumption of such intuitive immediacy, I suspect, that presses him toward dualism rather than toward exploring the possibilities of materialism without mechanical reduction (though perhaps Chalmers would call a materialism that embraces mechanical explanations for some kinds of facts and events and nonmechanical explanations for others a kind of "dualism"; see note 3). It is true, as he argues, that the mechanical explanations of science since Descartes and Newton have replaced senses of mysteries in the world with more or less satisfying "material" explanation. But it seems possible that one can still remain a naturalistic materialist without solely relying on mechanical or *physicalist* explanations. Chalmers consistently identifies materialism and a physicalist conception:

> There is also a natural tendency to believe that everything is physical and that consciousness must be physically explainable one way or another. . . . Once we accept that materialism is false, it becomes clear that the search for a physical X-factor [determining consciousness] is irrelevant; instead, we have to look for a "Y-factor," something *additional* to the physical facts that will help explain consciousness. (1996, 245)

In this argument, Chalmers presents a particularly narrow sense of fact and a narrow sense of materialism, very different from what

Wiener meant when he argued that "information is information, not matter or energy. No materialism which does not admit this can survive at the present day" (1961, 132). With this assertion, Wiener suggests that materialism can encompass wider modes of understanding and explanation than mechanical or physicalist reduction. He does so by suggesting that information itself is a material phenomenon (but not necessarily reducible to a physical phenomenon conceived in terms of fact). "Information," he writes, "is a name for the content of what is exchanged with the outer world as we adjust to it, and make our adjustment felt upon it" (1967, 26–27)—the very interactions George Levine and John Casti describe—and the example Wiener uses is Leibniz's examination of optics as opposed to Newton's mathematical physics because Leibnizian optics is related to Chalmers's sense of experience in ways that physics is not insofar as Leibniz emphasizes messages and communication. But before I turn to conceptions of information—such notions have recently multiplied in our intellectual lives—I want to spend a moment on the notion of fact at the heart of the reductive materialism Chalmers both embraces and rejects.

Intangible Materiality and Semiotic Facts

In *A History of the Modern Fact,* Poovey argues that our "modern" notion of *fact* arose in the seventeenth century. "Because most modern sciences in the West (including philosophy)," she writes, "position the category of the factual between the phenomenal world and systematic knowledge, the epistemological unit of the fact has registered the tension between the richness and variety embodied in concrete phenomena and the uniform, rule-governed order of humanly contrived systems. Because facts register this tension, in turn they have been susceptible to the two interpretations I have described" (1998, 1–2). The first of the two interpretations she presents understands facts, on the one hand, "as particulars, isolated from their contexts and immune from assumptions (or biases) implied by words like 'theory,' 'hypothesis,' and 'conjecture'"; and, on the other hand, a second interpretation understands facts as "evidence that has been gathered in the light of—and thus in some sense *for*—a theory or hypothesis . . . [so that] facts can never be isolated from contexts, nor can they be immune from the assumptions that inform theories" (1). In Poovey's history, fact is never simply the "physical facts" that Chalmers describes. (Nevertheless, her description does take in the "hard" problem of consciousness—its seeming phenomenal immediacy—as Chalmers describes it in relation to the systematicity of psychological behavior.) Rather, in her history, she argues that a "reformulation" of the notion of fact began "to displace the modern fact after about 1870." "This reformulation," she contends,

> which occurred at different moments in different disciplines, gradually elevated rule-governed, autonomous models over observed particulars. After the late nineteenth century, at least in the natural and social sciences, expert knowledge producers sought not to generate knowledge that was simultaneously true to nature and systematic but to *model* the *range of the normal* or sometimes simply to create the most sophisticated models from available data, often using mathematical formulas. As the units of such models, postmodern facts are not necessarily observed particulars; instead, as digital "bits" of information, the "phenomenological laws" of physics, or poststructuralist signifiers with no referent, they are themselves already modeled and thus exist at one remove from what the eye can see, although they are no less the units by which we make what counts as knowledge about our world. (1998, 3–4)[6]

Such facts disclose, or at least instantiate, the organization of matter; they remain material insofar as they are not essentially or qualitatively different from the physical, as they would be in the kind of dualism that Chalmers advocates. They, in fact, *constitute* the physical.

But they do so, as Poovey suggests, in ways that studies of information, phenomenology, and semiotics can help us understand. That is, rather than taking the immediately given as "brute" facts (as Chalmers does), these so-called postmodern disciplines attempt to find modes and moments of *mediacy* in relation to the given: combinations of what has been described, in relation to linguistics and semiotics, as a "phenomenological structuralism" (Holenstein 1976) that grasps together what Chalmers claims are completely distinct, phenomenal experience and determinate formalism. (They are also distinguished in Poovey's twofold definition of the "modern fact" before the postmodern moment she describes.) Such grasping together is clearest, I think, in relation to semiotic systems insofar as meaning is both *felt* and *understood*: meaningfulness (and even cognitive bewilderment) seems to strike us immediately (or at least "directly")—so that "logical properties," as Lévi-Strauss argues, are "manifested as directly as flavors or perfumes" (1975, 14)—and yet, like Poovey's notion of fact, they exist within or in relation to (semiotic) systems and within processing in neurological time even while they feel as if they are *immediately* apprehended.[7] The particular semiotic system that Poovey focuses on in her study of the early modern fact (she does not use the term *semiotic*) is the emergence of double-entry bookkeeping around the beginning of the sixteenth century, in which "the availability of a prototype of the modern fact in a familiar (but socially devalued) cultural practice like commerce enabled natural philosophers to explain what kind of knowledge they wanted to produce" (1998, 11). The genius of double-entry bookkeeping was its combination of "formal precision" and (at least the illusion of) "referential accuracy" (55; see also 108): "The double-entry fact—the numbers that appear in the ledger—[is] an intermediary between empirical events and the theoretical system that constitutes the site of general knowledge. As intermediaries, the double-entry numbers seem both to record transactions that actually occurred *and* to belong to (indeed, to emanate from) the double-entry system itself" (76–77).

I present Poovey's focus on double-entry bookkeeping in order to describe the "modern fact" because within economic accountings such facts include the odd category of *intangible assets,* from which I borrow the title of this chapter and book. This category allows for the *material* accounting of assets that are not *physical* or subject to the kinds of physicalist reductions that Chalmers describes. "Intangible assets" describes value that is somehow associated with a business but not located in a palpable item such as cash, notes, accounts receivable, inventories, fixtures, machines, or real estate; it describes business phenomena beyond the tangible matters-of-fact of mechanical materialism. Such intangible assets include the technical economic category of "goodwill." Originally, the term described what the OED defines as "the privilege, granted by the seller of a business to the purchaser, of trading as his recognized successor; the possession of a ready-formed 'connexion' of customers, considered as an element in the saleable value of a business, additional to the value of the plant, stock-in-trade, book-debts, etc." The OED cites the earliest usage in this sense in 1571; the latest usage it states occurs in 1876. A good example of this intangible asset can be seen in the way a physician sells his practice and, in so doing, introduces his patients to the new doctor and recommends her skill. By the time Thorstein Veblen wrote *The Theory of Business Enterprise* in 1904, the successor to *The Theory of the Leisure Class* in which he describes what we now would call the "cultural capital" of "conspicuous consumption," he describes a more pervasive sense of the intangible asset of goodwill. In *The Theory of Business Enterprise,* his example of goodwill is the value embodied in common stock. Veblen notes that "the question of 'stock watering,' 'overcapitalization,' and the like is scarcely pertinent in the case of a large industrial corporation financed as the modern situation demands. Under modern circumstances the common stock can scarcely fail to be all 'water,' unless in a small concern or under incompetent management. Nothing but 'water'—under the name of good-will—belongs in the common stock; whereas the preferred stock, which represents material equipment, is a debenture" (1932, 117–18). In this passage, Veblen transforms common stock into "reputation" rather than a company's "tangible" assets (which are what preferred stock represents). The phenomenon of widespread and primarily anonymous trading of common stocks, like other "postmodern" facts arising at the turn of the twentieth century,[8] is a far cry from the hands-on introduction of the new doctor to one's patients or of the new shop owner to one's neighbors: in many ways, the stock market is all reputation—messages, communication, and "information."

Veblen decries the social and human costs of the transformation of industrial capitalism into finance capitalism,[9] but he also describes, as does Poovey, the phenomenological and material effects of bookkeeping altogether. The intangible asset Poovey describes in double-entry bookkeeping—its goodwill—is what she describes as the "effect of accuracy" created by the semiotic precision of double-entry books that, in turn, creates the effect (or "felt sense") of honesty and, especially, the honesty of "total disclosure" (1998, 58), the "transparency" of its accounting (64):

> In numerical representation and double-entry bookkeeping, the merchant possessed devices designed to record economic transactions while making the writer seem to disappear. For this reason, the ability to "register" natural matters of fact "honestly"—that is, uncorrupted by preconceptions or theories—could be equated with the use of instruments that could generate the effect of transparency and therefore of impartiality, whether those instruments were merchants or the conventions of representation that merchants used. (116)

Such honesty is felt to be immediate, self-evident, simply a matter of fact. It is what the negotiation between empiricism and system creates along with its other, more palpable "assets": record, accounts, a sense of the value of enterprise measured in numbers.

The pervasive goodwill Veblen describes situates itself among the "postmodern facts" Poovey catalogs in science and culture in the early twentieth century. Her example is economics (which is why Veblen provides a fine example of the kinds of intangible facts she is describing), but it can be seen in notions of materialism that are implicit in the sciences as well. In the second chapter of *The ABC of Atoms* (1923), Bertrand Russell describes "The Periodic Law," focusing on the periodic table of elements articulated by Dmitri Mendeleeff (more commonly spelled Mendeleev) in the 1870s. The periodic table organizes chemical elements systematically and, as Russell says, "it has proved itself capable of predicting new elements which have subsequently been found" (20). It does so by allowing us to apprehend particular regularities in chemical properties by means of the ways that it organizes phenomena. Similarly, and virtually contemporaneously with Mendeleev's periodic table, Ferdinand de Saussure—in *his* systematic account of what have come to be called phonemes (and which might well be called the "elements" of language, analogous to the chemical elements the periodic table organizes)—was able to predict the existence of hitherto unknown phonemes based on the regularities of their systematic organization (see Culler 1976, 66). Thus, Russell writes,

> When once the periodic law had been discovered, it was found that a great many properties of elements were periodic. This gave a principle of arrangement of the elements, which in the immense majority of cases placed them in the order of their atomic weights, but in a few cases reversed this order on account of other properties. For example, argon, which is an inert gas, has the atomic weight 39.88, whereas potassium, which is an alkali, has the smaller atomic weight 39.10. Accordingly argon, in spite of its greater atomic weight, has to be placed before potassium, at the end of the third period, while potassium has to be put at the beginning of the fourth. It has been found that, when the order derived from the periodic law differs from that derived from the atomic weight, the order derived from the periodic law is much more important. (1923, 23)

The periodic table assigns an atomic number to every element, and as Russell says a bit later, "the atomic number is much more important than the atomic weight" (24–25; he also spells out the differences between the weight and number of atoms, 26–27).

In important ways, Russell is describing what Poovey calls the "postmodern fact," which she dates, as does Russell, about 1870. These "facts," as she says, "are not necessarily observed particulars" but rather are systematically "modeled and thus exist at one remove from what the eye can see, although they are no less the units by which we make what counts as knowledge about our world" (1998, 3–4). Such modeling, I think, is precisely what allows the possibility of conceiving of materialism beyond physicalism—to conceive of an *intangible* materialism—especially in relation to "nested" stages of complexity. This might become clearer if we look at Russell's *The Analysis of Matter,* published a few years after *The ABC of Atoms.* Toward the end of that book, he returns to periodicity, and specifically to Planck's constant, *h,* which, he argues, "only exists, or at any rate is only important, in the case of periodic processes, and it is a characteristic of one complete period" (1954, 365). "It is impossible," he concludes,

> to resist the view that *h* represents something of fundamental importance in the physical world, which, in turn, involves the conclusion that periodicity is an element in physical laws, and that one period of a periodic process must be treated as, in some sense, a unit. (365–66)

Such a "unit" seems hardly "physical" in any usual sense of the word (though Russell argues in *The Analysis of Matter* that matter is best conceived as an "event" rather than as substance [243–44]). Or rather, his analysis asks us to reconceive (or, to use Poovey's term, reformulate) physicality or at least materialism in relation to time—what Russell calls the relation of cause to interval. "The causal unit," he writes, "on relativity principles, should be expected to occupy a small region of space-time, not only of space; it should not therefore be instantaneous, as in pre-relativity dynamics" (366). Earlier, he had described "the relation of a steady event to a rhythm I conceive according to a musical analogy: that of a long note on the violin while a series of chords occurs repeatedly on the piano" (363). In relation to the "steady event" of matter, he argues, "all change should be discontinuous" (362). Still, such change creates periodicity, which he says, "perhaps fancifully," is "something faintly analogous as an accompaniment to every steady event. There are laws connecting the steady event with the rhythm; these are laws of harmony. There are laws regulating transactions; these are laws of counterpoint" (363).

The temporalization of matter suggests that Russell's conception of periodicity is akin to Wiener's conception of information: it is a process over time that also involves interaction and exchange.

It is not purely accidental that Russell turns to a musical analogy in his description of the dynamism of matter. Victor Zuckerkandl describes "three modes of being" in what he calls the three components of sense perception in his remarkable study of the philosophy of music, *Sound and Symbol.* He asks "how, without falling back upon the old belief in the world soul or in a God in nature, [are we] to conceive feelings outside of a consciousness, and a seeing, hearing, and touching of feelings?" (1969, 60). To answer, Zuckerkandl argues that "music shows us the way out" (60). "Tone sensations," he says,

> of course, are subject to the same law as all other sensations: they too are always colored by feeling. . . . Now, whether we interpret the emotional tone as something contributed by the hearer or as a quality of the tone itself, one thing is certain: the *musical tone* cannot be adequately described in terms of these two components, the physical and the psychic. Let us think of the Beethoven melody. . . . When it appears for the first time in the Ninth Symphony, it is played by the lower strings. The tones of the celli and double basses in this passage—especially by contrast with what has preceded—have a very definite emotional character: it could be called a character of solemn repose. The two components, then, are present—the physical, the acoustical tone, and the psychic, the emotional tone; but the *melody,* the *music,* as we know, is in neither of these. What we hear when we hear melody is simply not F sharp, G, A, etc., plus "solemn repose," tone plus emotion, physical plus psychic, but, with that and beyond it, a third thing, which belongs to neither the physical nor the psychic context: . . . a pure dynamism, tonal dynamic qualities. It is not *two* components, then, which make up musical tone, but *three.* (60)

The third component Zuckerkandl describes—the terms he uses, besides *dynamism,* are *force, equilibrium, tension, direction*—stands alongside the physical fact and the emotional (phenomenal) meaning. Later, he describes it as "the core of all that is manifest to the senses" (63); it is, as I note later in this chapter, Peirce's icon, standing with indexical fact and symbolic meaning. This phenomenon, which he calls the "external psychic," would then prove, he says, "to be something purely dynamic, not feeling but force—a force for which the physical would be as it were transparent, which would work through the physical without touching it." He claims that rather than the eye—whose visual apprehension "has such an important part in the construction of the world of material things"—the ear "is the organ particularly capable of perceiving the dynamic component of external events" (63). It might be interesting to speculate that the impossibility of visualizing Heisenberg's mathematical description of quantum mechanics that so "disgusted" Schrödinger could be countered by a sense of periodic dynamism that Russell describes, susceptible to the ear. John Casti (1995) studies at great length the role of observation in science, and especially the visualizations of model building. Might it be that physicalist, tangible materialism need not preclude an unvisualizable materialism "for which the physical would be transparent" without rendering such materialism ghostly? Later in this chapter, I suggest that such unvisualizable materialism might well be what we mean by "information."

Meaning, Interpretation, and Information

Russell's analogy and Zuckerkandl's analysis are, indeed, the kind of modeling—even postmodern modeling—that Poovey describes. And they return us to the significance of *meaning* with which I began. In *Semiotics and Language: An Analytical Dictionary,* A. J. Greimas and J. Courtés offer two definitions of *interpretation.* According to the "classical" concept of interpretation, they write, "every system of signs may be described in a formal way that does not take into account the content and is independent of possible 'interpretations' of these signs" (1982, 159). In this conception—which governs the mechanical formalism that Chalmers describes—"semantic interpretation" comes after and follows from abstract formalism, which is general and universal. This understanding of interpretation preserves our commonsense idea of the referential function of interpretation and sign systems more generally: the world to which language refers *preexists* the language that describes it[10] just as the necessities of formal logic preexist the phenomena that exemplify those logical necessities. Chalmers's analysis of consciousness is self-consciously logical in this sense. And it preserves, as well, the "instantaneous" sense of cause that Russell associates with prerelativity dynamics. Interpretation, then, consists of finding a language that *conforms* to preexisting *forms;* interpretation always come after the fact.

The second concept of interpretation, Greimas and Courtés argue, is completely different. Within this perspective—which they identify with Saussurian linguistics, Husserlian phenomenology, and Freudian psychoanalysis, all of which arose within the time period Poovey describes and which nicely comport with her sense of the modeling nature of postmodern facts—"interpretation is no longer a matter of attributing a given content to a form which would otherwise lack one; rather, it is a paraphrase which formulates in another fashion the equivalent content of a signifying element within a given semiotic system" (1982, 159). In this second understanding, *meaning* is an analogical paraphrase of something which *already* "signifies within a system of signification," a kind of translation built upon other, already-existing models, like Russell's fanciful analogy between music and periodicity.[11] In this conception of interpretation, the meaning or semantic content is not dependent on preexisting forms. Rather, both content and form—the meaning and its imbrication within a formal, causal nexus, which is to say the phenomenal and empirical "factuality" of meaning that Chalmers distinguishes as phenomenal experience and psychological causality—arise together, collaboratively and analogically. Later, I describe this as material formalism that grasps as a whole what Chalmers separates under the categories of experience and fact. It is what I called in *Analogical Thinking* "the *notation*—that is, the *recognition* and the disciplinary *inscription*—of meaning and value" (2000a, 108).[12] Such *notation*, I believe, functions like the systematics that Russell describes in the periodic table; it is the realization of a nonreductive materialism insofar as it does not posit, as Chalmers does, a "fundamental" dualism, nor does it reduce phenomena to abstract formalism. Moreover, as a form of *recognition,* it is fully embedded within the definition of information that Wiener presents.[13]

In his mathematicization of information, Claude Shannon focused on the sender of a message rather than on the receiver (who "recognizes" the message). Such a strategy, as Katherine Hayles has noted, "brackets semantics" (1999, 54). Donald MacKay suggested that such a definition was too narrow. Thus, in Hayles's account, MacKay proposed that Shannon was

> concerned with what he called "selective information," that is, information calculated by considering the selection of message elements from a set. But selective information alone is not enough; also required is another kind of information that he called "structural."

> Structural information indicates how selective information is to be understood; it is a message about how to interpret a message—that is, it is a metacommunication. (Hayles 1999, 55)[14]

The problem of the reception of a message is the crucial problem for classical science (as it is, as Zuckerkandl suggests, for music as well): while the sender of a message can always be universalized—and understood, as it is in narrative theory, as a transcendental subject, much in the way Shannon understands it[15]—the receiver of a message cannot quite so easily be universalized. In *The Human Use of Human Beings*, Wiener contrasts Leibnizian optics with Newtonian mechanics. Leibniz's optics, Wiener argues, like his "monads," above all is involved with "perception" (1967, 28) and—perhaps like the role of the observer in quantum physics[16]—it is involved with interaction and feedback.

Wiener describes information in terms of patterns of exchange. (Freud makes the same observation in 1904—in the midst of the emergence of Poovey's postmodern fact—when he argues in *Dora* that symptoms have to occur more than once in order to be symptoms [1997, 33–34]. Here too is a version of Russell's periodicity as well as Zuckerkandl's subtle description of the function of repetition in music [Zuckerkandl 1969, 212–23].) Such "feedback" exchanges characterize biological systems—and especially evolution—more fully than they do mechanistic systems. Moreover, they characterize semiotic systems in a striking way. As I argue in chapter 2, while physics and biology examine and describe the creation of order, semiotics systems *necessarily* entail possibilities of disorder. As Bruce Clarke notes,

> The mathematicization of information in communication theory established it as a scientific entity on a par with the physical concepts of matter and energy. The advent of information as a fundamental category also accompanied the relativization of matter and energy, which no longer possess the absolute existence implied by traditional laws of conservation. Both *can* be created and destroyed, but only (as quantified by Einstein's famous equation $e = mc^2$) by expending or producing the other. Thus a higher form of conservation law still determines, with relativistic adjustments, our understanding of the limits of material and energic phenomena; moreover, they are interconvertible only under what to us are extreme conditions. In contrast, information is inherently hyperbolic at all scales; more precisely, it

> has no proper scale. As with entropy, conservation laws do not apply. (2002, 31)

Entropy and the laws of thermodynamics, Percy Bridgman has observed, have "something more palpably verbal about them—they smell more of their human origins" (1941, 3; see also Campbell 1984, Pt. I). This is why the connection between information and physics—between the *phenomenology* of meaning and the *mechanics* of matter—are not so far apart as Chalmers suggests. In fact, there is a *structural* homology between them, rather than Chalmers's "natural dualism," so that materialism can encompass the *experience* of meanings—disorderly as they might be—as well as the orderly organizations of matter and energy.

Information and Materialism

In *Modernism and Time,* I offer an examination of Werner Heisenberg's philosophical explorations of quantum physics, which ends with a section titled "The Ontological Argument" (see Schleifer 2000b, 203–7). I suggest that Heisenberg is returning to the pre-Enlightenment conception of existence as a quality that, with others, can be attributed to an entity. In his discussion, Heisenberg describes the "geometrical properties" of atoms. "Anything that can be imagined and visualized," he writes in *Philosophical Problems of Nuclear Science,*

> cannot be indivisible. The indivisibility and homogeneity, in principle, of elementary particles makes it quite understandable that the mathematical forms of atomic theory can hardly be visualized. It would even seem unnatural if atoms lacked all the general qualities of matter like colour, smell, taste, tensile strength, and had yet retained geometrical properties. It is much more plausible to think that all these properties can be attributed to an atom only with the same reservations, and such reservations may also later enable us to relate space and matter more closely. The two concepts, atom and empty space, would then no longer stand side by side yet be completely independent of one another. (1952, 120)

In our usual understandings, the geometrical properties of location in time and space are not equated with the other properties Heisenberg lists here. Color, smell, taste, strength—the "*immediate and direct* physical properties" Heisenberg speaks of elsewhere (42), the "sense-properties" of matter (81)—are understood as attributes, detachable "parts," of some (positive) underlying whole. To speak of atoms "retaining" geometrical properties is to equate them with something seemingly outside of them, their location in space, not a part at all. Democritus, Heisenberg notes, deprived the atom of the qualities of color, smell, and taste, "and his atom is thus a rather abstract piece of matter." But, he goes on, Democritus "has left to the atom the quality of 'being,' of extension in space, of shape and motion." He has left these qualities, Heisenberg argues in *Physics and Philosophy,* because "it would have been difficult to speak about the atom at all if such qualities had been taken away from it" (1958, 69–70). In other words, in his account of quantum physics, Heisenberg reintroduces "being" as a quality or attribute rather than the positive physical substance—the *material* substance—to which other qualities can be attributed: he reintroduces the premises of Saint Anselm's ontological argument for the existence of God that Kant had seemingly demolished in the *Critique of Pure Reason* (earlier, Hume did so as well) when he argued that existence is not a predicate or an attribute of an entity in the same way that its color or taste is.

Implicit in Heisenberg is a particular, counterintuitive notion of matter and materialism that contradicts the Enlightenment modernist notions of matter, "the quality of 'being,' of extension in space, of shape and motion" that, as Heisenberg says, must be visualizable. (It is for this reason that Erwin Schrödinger said that the impossibility of picturing Heisenberg's mathematical description of quantum mechanics was "disgusting, even repugnant.") In any case, the conception of matter as essentially extensive is most fully identified with Descartes who, Harold Johnson argued, made "sheer geometrical extension the essence of material body" (1973, 193). Needless to say, such an assumption about the nature of matter makes sense coming from the inventor of analytical geometry and the founder, with Newton, of mathematical physics. For both Descartes and Newton, metaphysical force—emanating from God—animates essentially inert matter. A second, related Enlightenment conception of matter as essentially in motion, energized, inhabits nineteenth-century conceptions of materialism that base themselves on the second law of thermodynamics. This is an extension of mechanics, yet it comprehends matter as essentially organized in relation to energy and motion. Both conceptions of matter as the inert object of force or the fluid object of energy—Heisenberg's extension and motion—are based on conceptions of continuity that Ernst Cassirer locates at the heart of Enlightenment modernism (1951, 30). (Michel Foucault describes continuity as "the indispensable correlative of the founding function of the subject" [1972, 12], and Hugh Kenner, speaking of Joyce, calls continuity "the most pervasive idea of the [nineteenth] century" [1978, 49]). In other words, both the physics and metaphysics of force and the physics and metaphysics of energy participate in "one of the assumptions of classical Newtonian mechanics," that Paul Davies and J. R. Brown note in the introduction to their collection of interviews, *The Ghost in the Atom:* "that the properties of matter are continuously variable." This assumption, they note, "breaks down on the atomic scale" (1986a, 1). Such breakdown, I suspect, conditions Heisenberg's more or less unconscious reintroduction of the ontological argument into discussions of materialism.

Heisenberg suggests that comprehension of the material of the universe, the elementary particles he studies, breaks down into the unvisualizable "mathematical forms of atomic theory" (1952, 120) that inhabit the postmodern facts Poovey describes. Such an unvisualizable fact, perhaps, is best seen in the paradoxes of Niels Bohr's Copenhagen interpretation of the wave–particle complementarity of light. "The Copenhagen view of reality," Davies and Brown argue, is "decidedly odd. It means that, *on its own* an atom or electron or whatever cannot be said to 'exist' in the full, common-sense, notion of the word" (1986b, 24). That is, they continue, "it is not a *thing* 'out there,' existing in its own right" (24; Martin Heidegger's 1930s lectures, later published as *What Is a Thing?* [1967], also address the problem of material facticity). In another book Paul (P. C. W.) Davies and John Gribbin try to visualize the problem of the wave–particle duality in another way. "It is important," they write, "to resist the temptation to regard electron waves as waves of some material substance, like sound waves or water waves. The correct interpretation, proposed by Max Born in the 1920s, is that the waves are a measure of *probability.* One talks of electron waves in the same sense as crime waves" (1992, 208)—in which, they continue, "crime wave" designates the "greater likelihood" of a burglary in some particular places. David Bohm makes a similar suggestion when he redefines the wave aspects of matter as "a new kind of wave which I call 'active information'" (1986, 128). Better than the vague figure of "crime wave," Bohm offers an analogy for the wave–particle duality that strikes me as particularly useful in that it does not quite depend on visualization. Bohm notes that the pre-Enlightenment philosopher Nicholas of Cusa asserted that reality had an "enfolded structure" such that "eternity both enfolds and unfolds time." "Well," Bohm goes on, "it's the wave-particle duality: you may say that something can unfold either as a wave-like or a particle-like entity. The mathematics of quantum mechanics—if you look at it carefully—corresponds to this enfoldment. It's very similar to the mathematics of the hologram." The hologram, he says, is a very good example of this "enfolded order" because in it "a pattern is enfolded into the photographic plate, and when you shine light on it it's unfolded into a visible image. Each part of the photographic plate contains information about the whole" (122).

Like the crime waves of Davies and Gribbin, this description presents an intellectual counterpoint rather than a visualizable image. What is striking about this analogy is the implicit connection it offers between materialism and information. Holographs, above all, are mechanisms of information storage[17] just as, less explicitly, probabilities of crime, assertions of crime waves, are assertions of information. If information seems less tangible—less *material*—than force or energy, if it seems to be the organization of matter rather than matter itself, what linguists call the "form of the content" (Hjelmslev 1961), we should remember, as Davies and Brown suggest, there is a particular *formality* to concepts like force and energy as well. "Take the concept of energy, for example," they write:

> Energy is a purely abstract quantity, introduced into physics as a useful model with which we can short-cut complex calculations. You cannot see or touch energy, yet the word is now so much part of daily conversation that people think of energy as a tangible entity with an existence of its own. In reality, energy is merely part of a set of mathematical relationships that connect together observations of mechanical processes in a simple way. What Bohr's philosophy suggests is that words like electron, photon or atom should be regarded in the same way—as useful models that consolidate in our imagination what is actually only a set of mathematical relations connecting observations. (1986b, 26)

Energy lends itself to purely indexical modes of representation, just as Newtonian "force" lends itself to the mechanisms of iconic modes of representation.

In a moment, I examine the unvisualizable materialism Heisenberg describes in quantum physics in relation to Peirce's category of symbolic modes, which is to say in relation to information theory itself. But before I do so, I want to configure, as Wiener does, information with force and energy. "Just as the machine-inspired metaphors of the past have produced physical concepts like force and energy," Tom Siegfried has argued, so "the computer metaphor has driven the development of a new form of physical science. Information is more than a metaphor. Information is real. Information is physical" (2000, 55). Siegfried asserts that information "is always embodied in some physical representation, whether ink on paper, holes in punch cards, magnetic patterns on floppy disks, or the arrangement of atoms in DNA" (66). His argument about the physicality—the *materiality*—of information is striking. Following the work of Rolf Landauer, he argues that the materiality of information is embodied in the energy that information processing consumes. (In the course of his argument, he cites Ralph Merkle's contention that "between 5 and 10 percent of the electrical power produced in the [United States] feeds computers" [75].) Landauer demonstrates, Siegfried maintains, that in information processing, "the only unavoidable loss of energy is in erasing it" (74). Information's physical enfolding of energy, like energy's enfolding of mechanical forces, underwrites its materiality. "Erasing information," Siegfried writes, "requires the dissipation of energy, Landauer's principle tells us. It sounds deceptively simple. But its implications are immense. It is the concrete example of the connection between physical reality and the idea of information as measured in bits" (240–41).[18]

Information, as Jeremy Campbell defined many years ago, is "the forces of antichance" in the universe, the "non-random element" in our world that is based, as he says, "on chance, but not on accident." Thus, information, he says, informs "the world in novel ways" (1984, 12, 11). More recently, Katherine Hayles described informational structures that emerge "from the interplay between pattern and randomness" (1999, 30), and in a provocative intimation, she suggests that information is a powerful way to define the crucially important but remarkably vague—that is to say, unvisualizable—notion of "event" (see p. 53), a suggestion that is akin, I think, to Russell's contention that matter is best conceived as event rather than as substance (1954, 243–44).[19] This sounds a lot like the pragmatics of Peirce's sign–object–interpretant semiotics, a lot like the information theory of MacKay that Hayles describes, inflected toward the receiver of the message: "In this model," Hayles says, "information was not opposed to change; it was change" (1999, 63). The analyses of both Peirce and MacKay might be grasped—interpreted—as the pragmatics of dislocation such as we find in many attempts to account for the materiality of quantum phenomena. In one such account, John Wheeler, like Heisenberg, suggests the impossibility of visualizing phenomena that are dislocated from space and time, from extension and movement. "If we're ever going to find an element of nature that explains space and time," he notes, "we surely have to find something that is deeper than space and time—something that itself has no localization in space and time. The amazing feature of the elementary quantum phenomenon . . . is exactly this. It is indeed something of a pure knowledge-theoretical character, an atom of information which has no localization in between the point of entry and the point of registration" (quoted in Davies and Brown 1986a, 66).[20]

A similar phenomenon can be seen in what Peirce describes as the "virtuality" of signification that delocalizes entry and registration: Peirce's symbol is future oriented and for that reason unlocalizable. That is, the symbol—and semiosis in general—is dislocated in time. "The only way a sign can stand for any object, regardless of how complex or artificial," John Sheriff argues in relation to Peirce,

> is by referring to it through previous thought. Since we are conscious only of previous thoughts and the most finite thought covers a time, immediate consciousness of thought-signs is not possible. The thought present in the mind is meaningful only by virtue of its relation to subsequent interpretant-thoughts that give us information about the objects reflectively. Meaning then, "lies not in what is actually thought [immediately present], but in what this thought may be connected with in representation by subsequent thoughts; so that the meaning of a thought is altogether something virtual" (Peirce 1931–35, 5:289). (Sheriff 1994, 37; the parenthetical phrase is Sheriff's addition)

In this virtuality, Peirce is arguing against the immediacy of meaning and—insofar as meaning is *experienced*—the immediacy of experience that Chalmers takes as his starting point. Moreover, the virtuality of semiosis and information shows a distant formal similarity to the quantum materialism suggested in the discussions of Heisenberg and Wheeler.[21]

In *Philosophical Problems in Nuclear Physics,* Heisenberg notes that "the atom of modern physics shows a distant formal similarity to the $\sqrt{-1}$ in mathematics. Although elementary mathematics maintains that among the ordinary numbers no such square root exists, yet the most important mathematical propositions only achieve their simplest form on the introduction of this square root as a new symbol. Its justification thus rests in the propositions themselves. In a similar way, the experiences of present-day physics show us that atoms do not exist as simple material objects" (1952, 62). Yet, as I note in *Modernism and Time,* such elemental particles in present-day physics do not exist as *nonmaterial* objects either. Instead, in the microphysics of quantum mechanics, they exist *outside the alternative* between material and nonmaterial just as $\sqrt{-1}$ is neither positive nor negative, neither a number nor not a number (see 2000b, 199). The information of mathematics, like energy and force, encompasses and enfolds the semiotic pragmatics and metaphysical materialist pragmatics—metaphysical in the powerful literalist sense of the term that Casti employs (1995, 79; I examine this topic more fully in chapter 2). The "distant formal similarity" Heisenberg describes here between the physical atom of modern physics and mathematical symbols is congruent with Wheeler's "atom of information" and Davies and Brown's description of energy as "a pure abstract quantity" (1986b, 26): such comprehensions seem to participate, pragmatically, in material reality itself and suggest a kind of material formalism that constitutes the "reality" of force, energy, and information that may help us rethink our understanding of matter and materialism.

Connoisseurs of Chaos

In chapter 2, I argue for the disorderliness of semiotics starting from the powerful examination of the creation of order in physics and biology, respectively, articulated by Erwin Schrödinger in his important book *What Is Life?* That is, I argue that semiotics is a "negative science"—opposed to positivism and positive science—in its play between order and disorder (Hayles's "pattern and randomness"). As we shall see, striking examples of such negative science are the ethics of Emmanuel Levinas and the rhetoric of Elizabeth Bowen, specifically in relation to the human face. Levinas asserts that ethics is the first philosophy, and he figures—and more than figures, he grounds—that philosophy in the encounter of person to person in the epiphanic recognition of face. Such recognition, he suggests, is the manifestation of disorder in order, the epiphanic recognition of "a mysterious forsakenness" (1989, 83) in the orderly world of our biological life. "The nudity of a face," he writes elsewhere,

> is a bareness without any cultural ornament, an absolution, a detachment from its form in the midst of the production of its form. The face *enters* into our world from an absolutely foreign sphere, that is, precisely from an ab-solute, which in fact is the very name for ultimate strangeness. The signifyingness of a face in its abstractness is in the literal sense of the term extra-ordinary, outside of every order, every world. (1996, 53)

Behind the material orders of physics and biology, there exists the absolute strangeness of materiality altogether in the face of the human (semiotic?) penchant for discovering—of "grasping," in Saussure's figure (1983, 229)—order in the world.

Levinas takes up disorder, *faces* it (so to speak) in the very disorder of transforming noun to verb (as I do here), object to encounter (the detachment of form he mentions), in order to face the overwhelming meaninglessness of brute materialism. Saussure does the same in his "science" of the arbitrary nature of the sign, where even nothing, as he says—the very absence, and what Roland Barthes has called the "degree zero," of language (1970)—can be taken up to the ends of meaning: "It is not necessary to have any material sign in order to give expression to an idea; the language may be content simply to contrast something with nothing" (1983, 86). In the last chapter of the *Course in General Linguistics*—accidentally "last," since the book was compiled posthumously by his students—Saussure argues against a causal explanation of descriptive linguistics in favor of a kind of Darwinian chaos and accident, the purely accidental (or, as Stephen Jay Gould says, "contingent" [1989, 288]) nature of any particular language form. That is, he argues that the differences between and within language groups are arbitrary and that rather than necessary cause-and-effect relationships between earlier and later language forms, between what comes before and what comes after, there are, in language families, simply phenomenal relationships—something like what Wittgenstein calls "family resemblance"—which can be recognized as relational only retrospectively. Similarly, it is only retrospectively, after one "faces" another, that the *basis* of philosophy, as Levinas sees it, arises; in Bowen's remarkable *negative* discourse, it is only retrospectively that her character's face arises out of descriptions of disparate facts. Such retrospective comprehension—*timely* comprehension—destroys the abstract formal authority of origin. "No feature" of language, Saussure writes, "is permanent as of right: it survives only by chance" (1983, 227).[22]

By way of example, he describes the characteristics of the Proto-Indo-European language and notes that "it is clear that none of these characteristics [of the Proto-Indo-European language] has entirely been preserved in the various Indo-European languages, and some of them . . . nowhere survive" (228). At the end, he compares the changing Indo-European characteristics with seemingly permanent characteristics of the Semitic languages. Saussure argues that even these apparent counterexamples to his thesis of the arbitrary nature of language are themselves accidental rather than necessary, the result of chance: "purely phonetic modifications, due to blind evolution." "But," he continues, "the alternations resulting were grasped by the mind, which attached grammatical values to them and extended by analogy the models fortuitously supplied by phonetic evolution" (229). Here again, the mind grasps (I used the term *takes up,* but both figures point toward the grasping of hand and index) phonetic accidents retrospectively in order to make them function linguistically.

More striking, however, is the choice of Hebrew in his argument. As Hillis Miller has argued, "No choice of examples is innocent" (1986, 2), and they are not because it is through the *notation* of both linguistic phenomena and linguistic strategies—that is, through both *factual encounter* and *phenomenal apprehension*—that language creates its signifying effects, what linguists call "meaning-effects." That is, exemplifying meanings participate in Greimas's and Courtés's second definition of *interpretation,* where meaning is "taken up," chosen in an act of attention, yet seemingly found. Such meanings—or, more technically, such meaning-effects—are "constructed" (the phenomenal action of grasping) even when they appear to be simply a transparent medium for expressing underlying truths, facts in the world. The choice of Hebrew as an example of the nonpermanence of linguistic characteristics and its ordering as the final example of Saussure's *Course* especially lacks the "innocence" of a simple encounter with a brute fact precisely because throughout Western history Hebrew was understood to be the sacred language, the language of Adam and of the Old Testament, the oldest, purest, least "accidental" or "brutal" language. For its permanence to be a function of chance, simply phonetic accidents that the mind grasps in order to make function in a particular way, toward particular ends, transforms the world from one in which God is the necessary first cause to one in which phenomenal results, functions, effects, and not causes, rule. For Saussure, the accidental permanence of Semitic linguistic characteristics functions in the same way that Bowen (in my example in chapter 2) seizes upon the adjectival detail of the young woman's lips, the color of "the underside of a new mushroom," to effect a particular interpretation. It transforms the world into events as random and accidental as the very existence of the brute material universe itself.

The play between order and disorder is nested systematically in relation to the orderly world of seeming purposefulness—the fittings of Darwinian adaptations—and to the engagements with "squirming" brute fact in mechanical physics that replaces its "brutality" with mathematical intelligence. In chapter 2, I call this semiotic play between order and disorder a negative science; here I describe it as an *activity* of literature. In later chapters, I situate this science and these activities in relation to the material realities of voice, hand, and pain. But let me end this chapter with Wallace Stevens's powerful meditation on the play of order and disorder, "Connoisseur of Chaos," and Peirce's remarkable organization of this play.

> I
>
> A. A violent order is disorder; and
> B. A great disorder is an order. These
> Two things are one. (Pages of illustration.) [. . .]
>
> III
>
> After all the pretty contrast of life and death
> Proves that these opposite things partake of one,
> At least that was the theory, when bishops' books
> Resolved the world. We cannot go back to that.
> The squirming facts exceed the squamous mind,
> If one can say so. And yet relation appears,
> A small relation expanding like the shade
> Of a cloud on sand, a shape on the side of a hill. (1972, 166–67)

The squamous mind is covered in scales, yet before this meaning is the accident that Stevens found in the sensuous nature of the sound of "squamous" repeating that of "squirming," so that sense—which is the level of Peirce's icon, phenomenal experience isolated from systems of fact and meaning—creates the effect of mirroring meaning: Stevens grasps this accidental sound pattern so that fact and mind seem, somehow, in relation to one another. The negative science here is the organized disorder of accidental sound, of misapplied descriptors (the abstraction of "mind" fitted with reptilian skin), of the phenomenology of felt meaning.

> IV
> A. Well, an old order is a violent one.
> This proves nothing. Just one more truth, one more
> Element in the immense disorder of truths.
> B. It is April as I write. The wind
> Is blowing after days of constant rain.
> All this, of course, will come to summer soon.
> But suppose the disorder of truths should ever come
> To an order, most Plantagenet, most fixed . . .
> A great disorder is an order. Now, A
> And B are not like statuary, posed
> For a vista in the Louvre. They are things chalked
> On the sidewalk so that the pensive man may see.
> The pensive man . . . He sees that eagle float
> For which the intricate Alps are a single nest. (1972, 167–68)

Experience—and I would add the *felt meaning* of experience—is the starting point of David Chalmers's argument about the nonreductive conscious mind. It is also the starting point of Peirce's semiotics—embodied in his icon. The very accident of Stevens's figure of the intricate Alps as a single nest—figuring, as you shall see, a central image in chapter 2, which takes up the seeming intangible "nestings" of materialism—might also give rise to felt meaning (now or later), a sense of order in the disorder of the ephemeral "chalkings," so to speak, of meaning in the world.

Peircean Facts

This is part of the work of literature: to return us to the material sensuousness of meaning, to the *materiality* of meaning. Such materiality is embedded in Peirce's conception of the sign and semiotics, particularly in his category of "icon."[23] "My view," Peirce wrote early in his career, "is that there are three modes of being, and [I] hold that we can directly observe them in elements of whatever is at any time before the mind in any way. They are the being of positive qualitative possibility, the being of actual fact, and the being of law that will govern the future" (1931–35, 1:23). The "qualitative possibility" he describes is the sheer sensation of phenomena, the "redness" of experience before it is associated with an object or a meaning, the "force" or "dynamism" of the musical tone: the *icon,* he says excites "analogous sensations in the mind" (2:299). The "being of actual fact" is apprehended by the *indexical* (or indicative) nature of signs: the fact that signs "point" or refer to objects in the world. "No matter of fact," he writes, "can be stated without the use of some sign serving as an index" (2:305); an index, he says, stands "unequivocally for this or that existing thing" (2:531); "anything which focuses the attention is an index" (2:285). The periodic table is such an index, focusing the attention; so are gestures of the hand, and the April wind in Stevens's poem. The difference between icon and index, then, is related, I suspect, to Ludwig Wittgenstein's observation that "pointing to the shape" of an object is different from "pointing to [its] color" (2006, sec. 33). Peirce's third modality, "law that will govern the future," is the modality of symbolism, and it nicely comports with the Danish linguist Louis Hjelmslev's redefinition of meaning as "purport" (1961, 55): future-oriented signaling, marking (literally? figuratively?) periodicity itself. For Peirce, a symbol is "a sign which refers to the object that it denotes by virtue of a law, usually an association of general ideas, which operates to cause the symbol to be interpreted as referring to that object" (1931–35, 2:249). With this definition, the periodic table or the hand's gesture or Stevens's figure is also a symbol. And, in fact, all signs—and experience and understanding themselves, I contend—participate in each of these modalities, even if one of these aspects of signs seem to dominate in any particular case.[24]

Peirce's distinctions can also be understood in relation to more general functions of language. In his study of the evolution of language and music, Steven Mithen argues that human protolanguage probably included a significant mimetic element—an *iconic* element—embodied in what he calls throughout his study prelinguistic "vocal gestures." Mithen presents experimental work that challenges "one of the most fundamental claims of linguistics: that of the arbitrary link between an entity and its name." The work of the ethnobiologist

Brent Berlin, he argues, "has shown that the names used for animals are frequently not entirely arbitrary but reflect inherent properties of the animals concerned, including the sounds they make (onomatopoeias), their size, and the way they move" (2006, 171). In addition to this iconic aspect of language, Peirce argues for the indexical—or indicating—aspect of language found in deictics, those words that, like gestures (including vocal gestures), indicate particular locations in time and space (e.g., *here, there, now, then*). Finally, Peirce's symbol indicates signs that seem altogether arbitrary, including the combinations of signs embodied in sentences and—in what some have called "narrative grammar"—discourse beyond the confines of the sentence.

Index and icon correspond, I believe, to Chalmers's categories of isolable physical fact and experience, data and phenomena,[25] and the crucial question for Chalmers (and for Peirce as well) is what, if any, possible kinds of communication between fact and experience—and, indeed, among all three of Peirce's "modes of being"—exist. Unlike Saussure, whose *arbitrary* signs emphasize the negative and dark side of the Enlightenment project (much as Levinas emphasizes the dark side of "factuality"),[26] Peirce works to encompass the materiality of signs—the materiality of phenomenality itself—within the organization of his concept of signs. He does this in his larger project of logic, in which he encompasses the intangible materiality of contrary-to-fact discourse within logic itself under the category of abduction. Max H. Fisch argues that Peirce's "major single discovery was that what he at first called *hypothesis* and later *abduction* or *retroduction* [now often called the logic of discovery] is a distinct kind of argument, different both from deduction and from induction, and indispensable both in mathematics and in the sciences" (1981, 17).

With his description of the logic of abduction, Peirce offers a way of understanding the relationships among his modes of being. In induction, Peirce argues, "we conclude that facts, similar to observed facts, are true in cases not examined," while "by hypothesis [abduction], we conclude the existence of a fact quite different from anything observed" (1992, 143; the essay also appears in Peirce 1931–35, vol. 2). Induction, he concludes, "is reasoning from particulars to the general law; [abduction], from effect to cause. The former classifies, the latter explains" (143). For Peirce, the logic of abduction connects an empirical result and the *quite different* case: abduction, he writes, "is where we find some very curious circumstance, which would be explained by the supposition that it was a case of a certain general rule, and thereupon adopt that supposition" (135). The case is different from the result because the case cannot be observed but only hypothesized; it is, as Casti says, an interaction "with another system" (1995, 269). Moreover, abduction is a mode of reasoning that is thoroughly historical, that reasons, as Gould says, not by prediction but by means of "postdiction" (1986, 65), a version of Peirce's retroduction and abduction.[27] That is, abduction, faced with the "very curious circumstances" that occasion its reasoning altogether, offers the hypothesis that an empirical event constitutes a case. Because abduction aims at discovering that a particular fact or circumstance is a "case" of a more general rule—in the way that musical tones are grasped as melody—it is bound up in the temporal periodicity (or what Gould calls the "iterated pattern" [65]) of narrative: the existence of an event as a "case" of a more general proposition is homologous to the existence of an event as an "episode" in a narrative that presents itself as a "meaningful whole."

In his examination of Darwin's "historical" science, Gould makes retrospective nature of abduction clear by describing the ways that Darwin brought together, as Peirce claims abduction does, "types of evidence" that are numerous and diverse: Darwin was, Gould argues, "so keenly aware of both the strengths and limits of history" that he "argued that iterated pattern, based on types of evidence so numerous and so diverse that no other coordinating interpretation could stand—even though any item, taken separately, could not provide conclusive proof—must be the criterion for evolutionary influence." And Gould adds parenthetically that "the great philosopher of Science William Whewell had called this historical method 'consilience of inductions'" (65). Gould's use of Whewell's term for history/narrative is significantly different from E. O. Wilson's use of the term examined in the next chapter. Unlike Wilson, Gould sees consilience as, in Darwin's words, a procedure that "groups and explains phenomena" (cited in Gould 1986, 65). Such grouping and explanation is the work of Peirce's abduction. Peirce calls such grouping the "characters" or "characteristics" of phenomena.

In chapter 2, I argue that semiotics is defined by the fact that it presents—and as a science focuses on—the "meaningful whole," the wholeness I mention in the introduction, in which the meaning of a sentence or a paragraph or a narrative is not reducible to any one element or group of elements it contains.[28] In a similar (and homologous) fashion, explicit and implicit narratives transform simple, seemingly isolated phenomena into meaning; they create what Greimas describes as "the still very vague, yet necessary concept of the meaningful whole set forth by a message" (1983, 59). Such a meaningful whole is the overall sense or point of an assertion or a story—its "thought," its "aboutness"—the meaning we take away from it, the moral of the tale, and/or even its sense of overall genre. Chalmers works hard to abstract experience (the experience of redness, for example) from any such "sense or point," yet Peirce suggests that while the experience of redness is a "mode of being"—a "positive qualitative possibility" (1931–35, 1:23)—and can be analyzed as a separate modality of experience/cognition, as a *possibility,* it is always imbricated in cognition just as future-oriented lawfulness is part and parcel of the experiences of meaning. Moreover, such future-oriented lawfulness entails narrative understanding, which is necessarily intangible, contrary-to-fact insofar as it is a promise of meaning, *purported fact.* The sense or point of a narrative—or even of an assertion—calls up the law of genre: thus, we say *Hamlet, Oedipus,* and even the death of John Kennedy are "tragedies" despite their great differences. We make this judgment (or simply have a more or less unarticulated sense of their similarity) because each of these very different narratives of Hamlet, Oedipus, and Kennedy configures or "grasps" a series of events—intellectual promise, prominent political power, unforeseen yet recognizable violence—in a manner that conveys or provokes particular cognitive and emotional responses. In this way, a vague sense of a whole unified meaning emerges from the elements of narrative presented (see Vannatta, Schleifer, and Crow 2005, chapter 2, screens 12 and 13, for a version of this passage).[29] Even seemingly simple assertions (e.g., "I like Ike"; "the atomic number is much more important that the atomic weight" [Russell 1923, 24–25]) always arise within a *situation* that can be narrated, that has a "point" or a "sense" that encompasses such narration (see Bahktin 1987 for a fine analysis of this phenomena). Russell, for instance, as I understand him, asserts his description of the periodic table against simple notions of positivistic "fact."

Peirce also resists simple notions of positivistic fact, and his logic of abduction is therefore closely connected to the working of *narrative* understanding as opposed to the *logical* understanding of deduction and the *empirical* understanding of induction.[30] Narrative organizes disparate events into complex action in which an array of happenings becomes the sequence of action that can be apprehended—retrospectively and simultaneously—as a whole. In its most basic form, narrative presents an initial situation, a change or reversal of that situation, and, most importantly, *retrospective comprehension*—that is, the *notation*—of what has taken place. Narrative comprehension accomplishes what Paul Ricoeur has called the "synthesis of the heterogeneous" by organizing a series of events "into an intelligible whole, of a sort such that we can always ask what is the 'thought' of this story" (1984, ix, 65): it grasps a meaningful whole. That is, for narrative, the whole is more than the sum of its parts, more than simply a collection of "data" added together. Umberto Eco describes this process in terms of Peircean semiotics as the transformation of "a disconnected *series*" into "a coherent [textual] *sequence*" that allows us to recognize the "'aboutness' of the text which establishes a coherent relationship between different and still disconnected textual data" (1983, 213). And Peirce himself argues that "the essence of an induction is that it infers from one set of facts another set of similar facts, whereas hypothesis [or abduction] infers from facts of one kind to facts of another" (1992, 150).

The second order of fact pursued by the logic of abduction, Peirce suggests, is "very frequently a fact not capable of direct observation" (150).[31] Moreover, Peirce describes the narrative nature of abduction when he distinguishes abduction from induction in terms of the fact that induction begins with objects and facts, while abduction begins with the characters or characteristics of phenomena (for instance, the redness of red) which have to be configured within particular categories or frameworks in order to be realized as facts. (It is for this reason, I argue, that Chalmers's dualism too absolutely distinguishes between physical fact and the further fact of experience.) "Hypothesis [abduction]," Peirce writes, "has been called an induction of characters. . . . [and] characters are not susceptible of simple enumeration like objects; [rather], characters run in categories" (140). Such categories entail the law that will govern the future.

The functioning of such a law—Peirce's symbol—is clear in one of the many little narratives Peirce presents in "Deduction, Induction, and Hypothesis," a narrative that reads very much like the stories of Arthur Conan Doyle, Peirce's contemporary.

> A certain anonymous writing is upon a torn piece of paper. It is suspected that the author is a certain person. His desk, to which only he has had access, is searched, and in it is found a piece of paper, the torn edge of which exactly fits, in all its irregularities, that of the paper

> in question. It is a fair hypothetic inference that the suspected man was actually the author. The ground of this inference evidently is that two torn pieces of paper are extremely unlikely to fit together by accident. (1992, 139–40)

Later in this paragraph, Peirce argues that "if the hypothesis were nothing but an induction, all that we should be justified in concluding . . . would be that the two pieces of paper which matched in such irregularities as have been examined would be found to match in other, say slighter, irregularities. The inference from the shape of the paper to its ownership is precisely what distinguishes hypothesis [abduction] from induction, and makes it a bolder and more perilous step" (140). In this example, the key difference between induction and abduction is conditioned by the knowledge of another *kind* of fact, namely, the knowledge that only the suspect "has had access" to the desk. This fact—which Peirce mentions here only in passing and does not mention again in his argument—is different *in kind* from the evidence of the torn paper both because it is focused on a different object from the inductive conclusions about the paper and because it is not observable: like the prior historical existence of Napoleon that Peirce also presents as simply a hypothesis (135), it is a "fact" that cannot be empirically observed.

At the time of the advent of what Poovey calls postmodern facts, Peirce was developing the logic of such factuality. Moreover, he was doing so without recourse to reductionism and without recourse to dualism. Like Saussure's historical conjectures—and like Stevens's figuration—he bases his logic on the materiality of signs, including their "intangible" materiality, which cannot be observed yet which has (communicative and notational) power in the world. Such materiality, in Peirce, encompasses the three modes of being he describes: iconic possibilities of qualitative experience, gestures of indexical reference, and the law of symbolic meanings. In the last three chapters of *Intangible Materialism,* I examine these levels of comprehension: in the meaningless language of Tourette syndrome (that nevertheless participates in the laws of cultural symbols), in the functioning of hand and gesture (that promises the operations of language), and in the felt sense of meaningful pain (that is at once sensuous, an overwhelming indicative fact, and a modality of meaning). I use the modes of being Peirce describes to help delineate the activities of literature, its sensuousness, its constant recognition of our situation the world, and its creation and instantiation of laws of meaning for cognition and experience.

In addition, I examine the affect of literature, its narrative impulse, and the ways in which it creates the phenomenology of subjectivity—perhaps the very consciousness that Chalmers studies. Toward the end of his remarkable study of the evolution of music in human cultures, *The Singing Neanderthals,* Mithen interrupts his speculations about the power of music for our prehistoric ancestors—whose sense of musicality was much greater than ours, he argues—with a description of how, sometimes, we may have "an intense musical experience that may capture some of the richness that was commonplace to the Neanderthals." "One of my own experiences," he writes, "was watching my first ballet, when I was suddenly made aware of how the human body can both express emotion and tell a story through its movement" (2006, 245). These two activities of the human body are the focus of chapters 3 and 4, the emotion expressed, involuntarily, in Tourette syndrome, and the storytelling functions of gestures of our hands. In chapter 5, I examine something Mithen doesn't quite discuss in his speculative evolutionary history, the sense of complex subjectivity that is pronounced in our encounters with the experience of pain and which also arises in encounters with the affective narratives of literature. These chapters, I hope, present a strong sense of what I call intangible materialism and suggest as well the materiality of the phenomenology of literary experience.

2. The Faces of Consilience: Levels of Understanding and the Negative Science of Semiotics

> The only objects whose behavior is truly "simple" exist in our own world, at the macroscopic level. Classical science carefully chose its objects from this intermediate range. The first objects singled out by Newton—falling bodies, the pendulum, planetary motion—were simple. We know now, however, that this simplicity is not the hallmark of the fundamental: it cannot be attributed to the rest of the world.
>
> —*Ilya Prigogine and Isabelle Stengers,* Order out of Chaos

At the beginning of Stephen Gaukroger's masterful intellectual biography of Descartes, he tells a remarkable story. "Since the eighteenth century," he writes,

> there has been in circulation a curious story about Descartes. It is said that in later life he was always accompanied in his travels by a mechanical life-size female doll which, we are told by one source, he himself had constructed "to show that animals are only machines and have no souls." He had named the doll after his illegitimate daughter, Francine, and some versions of events have it that she was so lifelike that the two were indistinguishable. Descartes and the doll were evidently inseparable, and he is said to have slept with her encased in a trunk at his side. Once, during a crossing over the Holland Sea some

> time in the early 1640s, while Descartes was sleeping, the captain of the ship, suspicious about the contents of the trunk, stole into the cabin and opened it. To his horror he discovered the mechanical monstrosity, dragged her from the trunk and across the decks, and finally managed to throw her into the water. We are not told whether she put up a struggle. (1995, 1)

This remarkable story—which Gaukroger assures us is apocryphal, even though late in the nineteenth century Anatole France made it a theme of a novel—suggests the economy of Cartesian understanding, the basic equation he grasps between "macroscopic mechanical phenomena" and "microscopic mechanical processes" (70) within "a thoroughly mechanist cosmology which takes as its foundations a strictly mechanist conception of matter" (255).

I begin my discussion of the plurality of modes of understanding with both Descartes's dream of understanding the world through a strictly mechanist conception of matter and the narrative and scandal to which it gave rise because there has always been, I think, a sense that mechanical explanation leaves too much out in its aims at simplicity, accuracy, and generalizability. I begin also with E. O. Wilson's remarkable book, *Consilience: The Unity of Knowledge,* because he shares with Descartes a wonderful passion for comprehension, not for saving the appearances (as Descartes's scholastic teachers had hoped) but for saving the essences.[1] Wilson imagines himself the inheritor of Descartes's dream to account for all phenomena by "the linking of facts," as he says early in his book, "and of fact-based theory across disciplines to create a common groundwork of explanation" (1998, 8). In the last chapter of his book, Wilson again asserts his argument "that there is intrinsically only one class of explanation. It traverses the scales of space, time, and complexity to unite the disparate facts of the disciplines by consilience, the perception of a seamless web of cause and effect" (266). These sentences present Wilson's definition of *consilience:* a "single parsimonious system" (5) of "organized knowledge" (44) that is materialist, secular, and unified. Wilson is a biologist of some note, having studied the organization of ant communities throughout his career and having published, twenty-five years ago, a book entitled *Sociobiology* that argued for the possibility and necessity of describing biological phenomena without recourse to vague, "social" or "purposeful" explanations. In *Consilience,* he even describes the bodies of ants as "walking batteries of glands filled with semiotic compounds" (70)—which is, I should say, the way I have thought about myself since adolescence. The determinist–causal model of explanation presented in *Consilience,* I suspect, would have allowed Descartes to fulfill his dream and perhaps to achieve at least an intellectual companionship that seemed to elude him. (His daughter Francine died in 1640 and left him with what his contemporaries described as "the greatest sorrow that he had ever experienced in his life" [Gaukroger 1995, 353].)

I have great sympathy for Wilson's materialist project, just as I have great gratitude for the Enlightenment dream Descartes and his followers created. Wilson sees in the Enlightenment—and especially in Enlightenment science—assumptions "of a lawful material world, the intrinsic unity of knowledge, and the potential of indefinite human progress" (1998, 8), and a "vision," as he says, "of secular knowledge in the service of human rights and human progress, . . . the West's greatest contribution to civilization" (14). At the same time, he fails to note in any way the *cost* of this secular vision, what Bruno Latour calls the Enlightenment's "double task of domination and emancipation" (1993, 10). Still, what troubles me most about Wilson, just as what troubled the sailors about Descartes, is the mechanistic reductionism of his vision, his repeated assertion that *all* phenomena can be ultimately understood in terms of the "simple universal laws of physics" (1998, 55), which he later describes as "explanation . . . by reduction [that] can in some instances be achieved across all levels of organization and hence all branches of learning" (71).[2] Time and again in *Consilience,* he claims that "an organism is a machine, and the laws of physics and chemistry, most believe, are enough to do the job [of creating artificial life], given sufficient time and research funding" (91). Elsewhere, he argues "the brain is a machine" (96), and like the identity Descartes posits between macroscopic mechanical phenomena and microscopic mechanical processes, it is best conceived of—is best *understood*—by means of the "parsimony," the generalizability, and the minute accuracy (71) of reductive analysis.[3] Above all, such reduction conceives of "human action"—and indeed, the animate action of natural selection "all the way down to the molecular level" (128)—as "events of physical causation" (11). Thus, consilience, Wilson asserts, "holds that nature is organized by simple universal laws of physics to which all other laws and principles can eventually be reduced" (55), including "the physical basis of the thought process itself" (64): "All tangible phenomena," he says in conclusion, "from the birth of stars to the workings of social institutions, are based upon material processes that are ultimately reducible, however long and tortuous the sequences, to the laws of physics" (266).

Levels of Organization

My issue with Wilson—and, implicitly, with Cartesian reductionism—is not his materialism but his notion of "reduction" to mechanistic physical description as the ultimate mode of explanation. John H. Holland distinguishes this extreme version of reductionism, which "holds that all phenomena in the universe are reducible to the laws of physics," with "a more cautious" position, which states "that all phenomena are *constrained* by the laws of physics" (1998, 188; see also Hayles 1991). Later, when I turn to the "level" of language and semiotics, I return to this issue of the negation of "constraint" in understanding. Stephen Shapin offers another version of this caution when he argues that "an account of the action of someone waving good-bye is not adequately given by detailing the muscular movements involved" (1996, 119); Jared Diamond does as well when he distinguishes between the "proximate explanation" of the skunk's smell in terms of chemical compounds with certain particular molecular structure and the "ultimate explanation" in terms evolutionary adaptiveness (1992, 124–25). And Ilya Prigogine and Isabelle Stengers offer still another example in their description of order through fluctuations. "'Order through fluctuations' models," they write, "introduce an unstable world where small causes can have large effects, but this world is not arbitrary. On the contrary, the reasons for the amplification of a small event are a legitimate matter for rational inquiry. Fluctuations do not cause the transformation of a system's activity. Obviously, to use an image inspired by Maxwell, the match is responsible for the forest fire, but reference to a match does not suffice to understand the fire. Moreover, the fact that a fluctuation evades control does not mean that we cannot locate the reasons for the instability its amplification causes" (1984, 206–07). (These three examples focus on semiotics, evolutionary biology, and physics.) Similarly, a mechanistic description of Francine—even her mechanical reproduction—can hardly explain the curious contradiction in the narrative with which I began that alternatively takes the doll to be indistinguishable from Francine and a mechanical monstrosity.

Wilson, however, consistently assumes the extreme position.

Throughout *Consilience,* he repeatedly asserts that by means of reductions to physical laws "all [phenomenal] facts will prove consilient. The explanations can be joined in space from molecule to ecosystem, and in time from microsecond to millennium." "With consilient explanation," he goes on,

> the units at different levels of biological organization can be reassembled. Among them will be whole plants and animals as we normally see them—not as collections of molecules in biochemical time, too small and fast-changing to be visible to the unaided eye, not as whole populations living in the slow motion of ecological time, but as individual plants and animals confined to the sliver of organismic time where human consciousness, being organismic itself, is forced to exist. (1998, 236–37)

Here is Descartes's dream with a vengeance, a whole made up—mechanically—of the sum of its parts.

The key term here is *levels,* for it is Wilson's implicit argument that the levels of organization—the physical/molecular level, the biological/environmental level, and the level of human meanings and understanding—are continuous and that movements across them are easy and reversible. This is clear in the very definition of consilience he offers. Wilson adopts the term from William Whewell's 1840 text, *The Philosophy of the Inductive Sciences,* and defines it—and in a crucial way *redefines* it—early in the book as

> a "jumping together" of knowledge by the linking of facts and fact-based theory across disciplines to create a common groundwork of explanation. [Whewell] said, "The Consilience of Inductions takes place when an Induction, obtained from one class of facts, coincides with an Induction, obtained from another different class. This Consilience is a test of the truth of the Theory in which it occurs." (1998, 8)

Good examples of Whewell's consilience—examples that took my breath away when I first encountered them—might be the efficacy of Planck's constant within different areas of inquiry (e.g., the integration of spectrum analysis and subatomic dynamics), or Darwin's connection of contingency and science (e.g., the integration of system and history), or even the general alternations of constative and performative messages in Freud, Austin, and Derrida (e.g., of the meanings of words and of actions). In *Consilience,* however, Wilson never returns to this definition nor uses the term in ways that conform to Whewell's conception of different classes of fact.

Stephen J. Gould offers a strong, *historical* description of Whewell's work that is at odds with Wilson's description.[4] "The great nineteenth-century philosopher of science William Whewell," he writes, "devised the word consilience, meaning 'jumping together,' to designate the confidence gained when many independent sources 'conspire' to indicate a particular historical pattern. He called the strategy of coordinating disparate results from multifarious sources consilience of induction" (1989, 282). Gould describes Darwin as the "greatest of all historical scientists" because he "explored a variety of modes for historical explanation, each appropriate for differing density of preserved information" (282). Later, Gould describes Darwin's consilience this way:

> Whenever he introduces a major subject, Darwin fires a volley of disparate facts, all related to the argument at hand—usually the claim that a particular phenomenon originated as a product of history. This style of organization virtually guarantees that Whewell's "consilience of inductions" must become the standard method of the *Origin*. Darwin's greatest intellectual strength lay in his ability to forge connections and perceive webs of implication (that more conventional thinking in linear order might miss). When Darwin could not cite direct evidence for actual stages in an evolutionary sequence, he relied upon consilience—and sunk enough roots in enough directions to provide adequate support for a single sturdy trunk of explanation. (2002, 109)

This narration of consilience is very different from Wilson's description of a "common groundwork of explanation," where the ground is the kind of radical reductionism that he describes, a whole made up of the sum its parts. Gould's consilience is more like Ricoeur's "configurational" modality of explanation examined in chapter 1, in which a consilience of facts forges webs of connection *across* different classes of facts.[5]

As I noted in chapter 1, Charles Sanders Peirce explicitly attends to the conception of different classes of facts in his discussions of hypothesis formation or the logic of abduction. Hypothesis formation or "abduction," Peirce wrote, "has been called an induction of characters . . . [and] characters are not susceptible of simple enumeration like objects; [rather], characters run in categories" (1992, 140). The "characters" or "characteristics" of phenomena might well be called their attributes or Cartesian and Lockean "secondary" properties (the color of the beans is Peirce's example in "Deduction, Induction, and Hypothesis" [1992], as opposed to the number of the beans with which deduction begins)—phenomena that troubled early materialists to the point of simply excluding them from consideration (as the logical positivists seemed to exclude the negative science of semiotics from their considerations). In other words, Peirce's abduction attempts to connect different *kinds* of facts, results and cases. Fritjof Capra addresses these differences in terms of "levels of complexity" (1997, 28), and throughout his work, he argues that "the key to a comprehensive theory of living systems lies in the synthesis of . . . two very different approaches, the study of substance (or structure) and the study of form (or pattern). In the study of structure we measure and weigh things. Patterns, however, cannot be measured or weighed; they must be mapped. To understand a pattern, we must map a configuration of relationships. In other words, structure involves quantities, while pattern involves qualities" (81). Later, he notes that "the difference between relationships among static components and relationships among processes [is] a key distinction between physical and biological phenomena. Since the processes in a biological phenomenon involve components, it is always possible to abstract from them a description of those components in purely physical terms" (164). Such abstraction, he suggests, fails to acknowledge the difference in the "class" and qualities of facts—different levels of comprehension. For Wilson, however, there is only a single, self-evident class of positive fact, and thus he redefines Whewell's term: in Wilson's science there is little trouble taken to examine—and, in his book, virtually no attention paid to—the issue of traffic back and forth across levels of phenomena, classes of fact.

J. T. Fraser describes what such trouble might look like in his description of levels of organization as "hierarchically nested." "The grouping of structures along hierarchically nested levels of increasing complexity," Fraser writes, "produces stability points. They make it possible for the next level of organization to emerge, because the lower levels do not get easily undone. In all cases, however, the issue is not one of simple aggregation but aggregation with communication" (1982, 24).[6] Fraser is studying different phenomena of time—his book is called *The Genesis and Evolution of Time*—but a "consilient" description of levels can be seen in Holland's description of the "emergence" of properties and phenomena that are not easily predictable from the initial conditions, "emergence in complex adaption systems—ant colonies, networks of neurons, the immune system, the Internet, and the global economy, to name a few—where the behavior of the whole is much more complex than the behavior of the parts" (1998, 2). Holland's chief examples are board games (checkers), computer playing and "learning" of board-game strategies, neural networks, and—like Wilson's and Douglas Hofstadter's examples—ant colonies. He argues that

> at each level of observation the persistent combinations of the previous level constrain what emerges at the next level. This kind of interlocking hierarchy is one of the central features of the scientific endeavor. . . . It will lead us into, and out of, the thorny thicket known as *reductionism*—roughly, the idea that we can reduce explanations to the interactions of simple parts. Because we are dealing with emergence in rule-governed systems, reduction has much to do with our exploration. . . . Here we see that emergence in rule-governed systems comes close to being the obverse of reduction. (8)

Such an "obverse of reduction," he says near the end of his study, takes place when we "move the description 'up a level,' replacing what may be difficult or even infeasible calculations from the first principles. These regularities still satisfy the constraints of the underlying microlaws, but they usually involve additional assumptions" (200–201).

Douglas Hofstadter is more precise—or at least more concrete—about the phenomena of nested hierarchies Fraser presents. In describing ant colonies, he suggests there is a certain kind of reductionism that cannot see the "macroscopic" forest for the "microscopic" trees.[7] More precisely in relation to gases, he argues that microscopic description "requires the specification of the position and velocity of every single component molecule," while the macroscopic description "requires only the specification of three new quantities: temperature, pressure, and volume." These parameters, he notes, combine in a "simple mathematical relationship . . . [that forms] a law which depends on, yet is independent of, the lower-level phenomena." "It is important to realize," Hofstadter concludes,

> that the high-level laws cannot be stated in the vocabulary of the low-level description. "Pressure" and "temperature" are new terms which

> experience with the low level alone cannot convey. We humans perceive temperature and pressure directly; that is how we are built, so that it is not amazing that we should have found this law. But creatures which knew gases only as theoretical mathematical constructs would have to have an ability to synthesize new concepts, if they were to discover this law. (1979, 308)

The synthesis of new concepts and new vocabularies is dependent on the possibility of organizing experience into levels that are *different in kind*—in Whewell's different "classes of fact," Peirce's different "characters" of fact—even if these levels are negatively constrained by one another. Pressure, like color or temperature, is not an appropriate category for understanding gas at the molecular level.[8]

The negative aspect of such differences are more clearly pronounced in Robert Kaplan's study of the "natural history" of the number zero. This "negative aspect" is crucial in comprehending the *material* work of hierarchical understanding. Kaplan describes different classes of fact in describing how "the uncomfortable gap between numbers, which stood for things, and zero, which didn't, would narrow as the focus shifted from what they were to how they behaved" (1999, 75). (Pressure, color, temperature, I might add, can all be understood as the *behavior* rather than the *state* of phenomena. The same is true for Wiener's "feedback" systems and, I argue, for semiotics in general.) The result of the narrowing of the difference between zero and other numbers, Kaplan says, was the progressive enlargement of vocabulary—in abstract words like "three" or "temperature"—as "the rabble of things" such as shoes or molecules and their position and velocity disappear: "Like zero," Kaplan writes, "numbers were becoming invisible: no longer descriptive of objects but objects—rarefied objects—themselves. 'Three' was once like 'small': it could modify shoes and ships or sealing wax. Now it had detached itself so far from the rabble of things that instead those ephemera participated briefly in its permanence" (75).

There is no such difference in kind, no disappearances replaced by new vocabularies, in Wilson's presentation of the continuous organization or what he calls the "reassembly" of different levels of biological organization; neither is there appropriate attention to difference in kind in the easy interchangeability of the doll's life-likeness and its mechanical monstrosity in the story of Descartes and "Francine." The models of organization Fraser, Holland, and Hofstadter describe and their levels of organization—which Mary Poovey might describe as "autonomous models [taking precedence] over observed particulars" (1998, 3)—allow for an understanding of phenomena that distinguishes Francine from her lifelike similacrum in ways that Wilson's Cartesian mechanism cannot, even as Wittgenstein's child talks to a "picture-face" (2006, 166): in terms of breathing, growth, and Descartes's own powerful category of speech. In the *Discourse on Method,* Descartes presents a mechanistic description of somatic life in which there is no room for interaction with the environment or the feedback of information theory. "We can certainly conceive," Descartes writes,

> of a machine so constructed that it utters words, and even utters words which correspond to bodily actions causing a change in its organs (e.g., if you touch it in one spot it asks what you want of it, if you touch it in another it cries out that you are hurting it, and so on). But it is not conceivable that such a machine should produce different arrangements of words so as to give an appropriately meaningful answer to whatever is said in its presence, as the dullest of men can do. (1985, 140)

Meaningful answers, as Descartes argues here, and even the "three strokes of luck in the evolutionary arena . . . . seemingly developed beyond their survival needs," which Wilson describes early in *Consilience* (1998, 48) in relation to the remarkable knowledge achieved by the natural sciences, cannot be adequately described in terms of the continuities of Cartesian equations of macroscopic and microscopic processes.

The models of hierarchically nested levels of organization described in Fraser's analysis of time, Holland's analysis of the emergent properties of systems, and Hofstadter's analysis of the contrapuntal references and self-references of codes, rather than the simple aggregation of Wilson's seemingly continuous levels, allow for a fuller accounting of a complex materialism than is possible through either the "strokes of luck" Wilson describes or Descartes's religion. In these models, Fraser argues, "hierarchical control is seen as arising from internal constraints that force the elements into collective behavior, independent of any detail of the behavior of the elements themselves" (1982, 24–25); similarly, Holland notes that "emergent behavior occurs without direction by a central executive" (1998, 6). That is, elements of one level combine to create the more complex basic elements of the next level that, on that new level, function as a simple "basic," or what Holland describes as "primitive," functions. In this way, the complex, emergent function is treated, on a higher level, "as if it were a function added to the set of primitive [simple] functions" (192). Prigogine and Stengers offer a description of the irreducibility of these levels in their discussion of Niels Bohr's conception of complementarity. "No single theoretical language," they write,

> articulating the variables to which a well-defined value can be attributed can exhaust the physical content of a system. Various possible languages and points of view about the system may be complementary. They all deal with the same reality, but it is impossible to reduce them to one single description. The irreducible plurality of perspectives on the same reality expresses the impossibility of a divine point of view from which the whole reality is visible. . . .
>
> The real lesson to be learned from the principle of complementarity, a lesson that can perhaps be transferred to other fields of knowledge, consists in emphasizing the wealth of reality, which overflows any single language, any single logical structure. (1984, 225)

From this vantage, the levels of description of phenomena—which is to say, the levels by which phenomena are comprehended conceptually and also in fact—do not allow for the easy traffic across levels that Wilson assumes.

On the level of natural selection, for instance, biology creates a second nature, so to speak, of biological, social, and even cultural environments composed of but different from the positive materialism of the physical world. Thus, speaking of cognition—a crucial theme in Wilson's discussion—Clifford Geertz argues that "the human brain is thoroughly dependent upon cultural resources for its very operation; and those resources are, consequently, not adjuncts to, but constituents of, mental activity" (1973, 76) in such a way that the developing nervous system adapted to the cultural as well as to the physical environment. "Culture," he concludes, "rather than being added on, so to speak, to a finished or virtually finished animal, was ingredient, and centrally ingredient, in the production of that animal itself" (47; see also Deacon's detailed study of "the co-evolution of language and the brain" [1998] and the "neural Darwinism" that Gerald Edelman describes in the development and functioning of the brain [2005]).[9] For this reason, a reductive mechanical explanation of simple accumulation in accounting for the "natural selection" of cognition is problematic: "The standard procedure," Geertz says, "of treating biological, social, and cultural parameters serially—the first being taken as primary to the second and the second to the third—is ill-advised" (47). Instead, the relationships among these parameters are nonlinear: as Holland notes more generally, "Emergence is above all a product of coupled, context-dependent interactions. Technically these interactions, and the resulting system, are *nonlinear:* The behavior of the overall system *cannot* be obtained by *summing* the behaviors of its constituent parts" (1998, 122).

The mechanism of the nonlinear hierarchical nestings of levels is even clearer in language and semiotics than it is in natural selection: in language, phonemes combine to create morphemes, which combine to create words; words, in turn, combine to create sentences. Sentences are made up of phonemes only in the most trivial sense: rather, sentences are combinations of words. Moreover, such combinations are not accumulative: the meaning of a sentence is not the accumulated meanings of its individual words. Rather, in language—but also even in the positive materialism of natural selection—the whole is greater than the sum of its parts. Such wholes, however, are not nonmaterial, neither unaccountable "luck" nor the design of some "central executive" of superadded instruction. Rather, they are the result of the nested hierarchies of material phenomena, what Hofstadter describes—in another term that encompasses nonreductive materiality arising out of levels of organization—as "epiphenomena," *"a visible consequence of the overall system of organization"* (1979, 308) that is not, "positively," located within any *part* that phenomena, yet also not simply a ghost in the machine. (It seems, rather, a kind of intangible asset.) This is clear in the Peircean modification of the "double articulation" of language that Saussure describes, which I take up in chapters 4 and 5. Saussure described language as articulated in both meaning (the "signified" of language) and "sound-images" (the "signifiers" of language), yet he defined the signifiers as phenomenologically apprehended rather than ("physical") substances in the world. Peirce's tripartite description of signification—arguing that signs are at once *iconic,* participating in the sensuous qualities of phenomena; *indexical,* pointing toward "facts" in the world; and *symbolic,* participating in interpersonal meanings—allows us to reconceive Saussure's signifier as at once iconic and indexical in such a way that a material conception of the signifier becomes both necessary and clear. (Even Saussure offers in *Course in General Linguistics* a diagram of the mouth that seems analogous to the embodiment of information "in some physical representation, whether ink on paper, holes in punch cards, magnetic patterns on floppy disks, or the arrangement of atoms in DNA," which Tom Siegfried notes [2000, 66].) But even more than the double articulation of sentences, the central problem for linguistic semiotics—namely, the problem of paragraphs and discourse beyond the sentence—more generally calls for the examination of the consequences of the overall system of organization that does not simply talk about a nonmaterial "ghost" or "spirit" of discourse. In this situation, negative science and its vocabularies of changing parts and whole become imperative.

Levels of Orderliness

Here, then, is my thesis: we can talk usefully, scientifically, and above all, materially about three levels of understanding: the positive science of physics, the "environmental" sciences (or context-dependent interactions) of biology and natural selection, and what I call—paralleling the positivism of physics (and of logical positivism, which Wilson calls "the most valiant concerted effort ever mounted by modern philosophers" [1998, 63–64])—the "negative science" of semiotics.[10] These levels present, in Capra's description of modes of scientific explanation, "different levels of complexity with different kinds of laws operating at each level. . . . At each level of complexity, the observed phenomena exhibit properties that do not exist at the lower level. For example, the concept of temperature, which is central to thermodynamics, is meaningless at the level of individual atoms where the laws of quantum theory operate. Similarly, the ['secondary quality' of the] taste of sugar is not present in the carbon, hydrogen, and oxygen atoms that constitute its components" (1997, 28).

I have presented examples of phenomena of language and of two modes of cognition—both the "interlocking hierarchy" that Holland describes as "one of the central features of the scientific endeavor" (1998, 8) and the work in artificial intelligence both he and Hofstadter examine—to describe Fraser's levels, but Fraser himself offers an extended example of the phenomenon of nested hierarchy from biology. "The hierarchical structuring of biological systems," he writes, "is a familiar fact. It has been worked out in detail by J. H. Woodger who . . . define[d] its concepts within the study of organismic biology. He noted that a member of any level may always be regarded from three points of view: from that of its own membership, from that of a member of a higher level, and from that of the next lower level. He also stressed the necessarily different kinds of lawfulness of the different levels and maintained that the laws determining the behavior of the parts of a given level of organization leave undetermined some aspects of the behavior of the parts of the higher level" (1982, 24; see also Woodger 1930 and Capra 1997, 28). This description of a biological example of nested levels, in addition to the examples I have presented, can be supplemented by the simple level of mechanistic physics, which, needless to say, I do not want to abandon.

Fraser's description of nested levels of phenomena—like the conception of "emergent" organization he pursues along with Hofstadter and Holland—is closely related to the framework Erwin Schrödinger articulates in his influential essay, *What Is Life?* in which he sets forth a description of biology from the point of view of a physicist. This small book, published near the end of World War II, is often taken to be the beginnings of microbiology; it offered to many physicists, disillusioned by the atomic bombs, the possibility of work in the biological sciences (Moore 1992, 403).[11] At the very beginning of that essay, Schrödinger argues that "the arrangements of the atoms in the most vital parts of an organism and the interplay of these arrangements differ in a fundamental way from all those arrangements of atoms which physicists and chemists have hitherto made the object of their experimental and theoretical research" (1992, 4). Toward the end of the essay, he puts this more forcefully: "From all we have learnt about the structure of living matter, we must be prepared to find it working in a manner that cannot be reduced to the ordinary laws of physics" (76). From the "picture of hereditary substance" available to him (writing in 1943), Schrödinger asserts, "It emerges that living matter, while not eluding the 'laws of physics' as established up to date, is likely to involve 'other laws of physics' hitherto unknown, which, however, once they have been revealed, will form just as integral a part of this science as the former" (1992, 68).[12] The great distinction Schrödinger makes between the laws of physics and the laws governing life processes is that the "statistical mechanism" of the laws of physics produces "order from disorder," while the laws governing life processes produce "order from order" (80). Peirce, a generation earlier, had conceived of this distinction in more general terms; as David Depew has argued, he "was a key participant in what historians of science now recognize as a Second Scientific Revolution, the probability and statistical revolution that forms the base on which twentieth-century science has been erected. It was Peirce, in fact, who first drove through to the nonmechanistic conclusions that his real peers, Maxwell, Boltzmann, and Gibbs, sought to avoid" (1994, 124). The "Second Scientific Revolution," as Poovey noted, entailed the "reformulation" of the notion of fact that began "to displace the modern fact after about 1870" (1998, 3). In this chapter, I describe such nonmechanistic conclusions as the negative science of semiotics Peirce helped initiate.

In his argument, Schrödinger surveys the levels I am presenting in

1. his examination of the difference between the periodic solids of crystals and the aperiodic solids of chromosomes and biological molecules more generally (1992, 59–61);
2. his definition of life as "'doing something,' moving, exchanging material with its environment . . . for a much longer period than we would expect an inanimate piece of matter to 'keep going' under similar circumstances" (69);
3. his remarkable conception of "negative entropy" that claims an organism's exchange with the environment exchanges neither matter nor energy but rather "orderliness" and what I would call organization, arrangement, and as I suggest in chapter 1, information.[13]

An organism, Schrödinger argues, "drinks orderliness" from "a suitable environment" (77), but such orderliness is not the *positive property* that Wilson suggests it is in his repeated defense of positivism in *Consilience* (1998, 61–65). Wilson even goes so far as to assert that "only observation can disclose the periodic table" (63), even though, as Russell argued in 1923, the genius of the periodic table is the way that its arrangement of elements—marked most notably in those cases in which the atomic number is different from the atomic weight—offers insight by means of semiotic arrangement rather than perceptual measurement. Thus, Russell writes, "it has been found that, when the order derived from the periodic law differs from that derived from the atomic weight, the order derived from the periodic law is much more important" (1923, 23). Rather, such orderliness creates the possibility of the "difference of property," which Schrödinger describes as "really the fundamental concept rather than property itself" (1992, 29) in a semiotic description remarkably similar to Saussure's conception of the sign system made up of differences.[14]

In *What Is Life?* Schrödinger's discussion of the atomic structure of solids and gases subject to the "laws of physics," his definition of life in "its internal life and its interplay with the external world" (18), and his "negative" descriptions of properties and order (the negation of entropy) present three levels of explanation that Wilson examines as continuous: the mechanics of physical science, the interactions of natural selection and of the biological and environmental sciences more generally, and the arbitrary and future-oriented substitutions of semiotics. Discussing "gene-culture coevolution," Wilson argues that the

> early stage of epigenesis, consisting of a series of sequential physico-chemical reactions, culminates in the self-assembly of the sensory system and brain. Only then, when the organism is completed, does mental activity appear as an emergent process. The brain is a product of the very highest levels of biological order, which are constrained by epigenetic rules implicit in the organism's anatomy and physiology. Working in a chaotic flood of environmental stimuli, it sees and listens, learns, plans its own future. (1992, 165)

Wilson is describing three levels of organization as if commerce among them is continuous even as he narrates his description, much as Holland, Fraser, and Schrödinger do, in terms of discrete levels of emergence and consolidation.

These levels offer different modes of organization: if physics creates order from disorder, and the natural selections of biology and environmental science—and also the "persistent" selections of Holland's and Hofstadter's cognitive sciences—creates order from order, then semiotics creates disorder from order. The relation between the first two levels I am describing—the level of physics and the level of biology—can be seen in three related discussions in Holland, in Hofstadter, and in semiotics more generally. In his discussion of emergence, Holland describes nonpurposive "action at a distance" accomplished by "signals" (1998, 145–46); in his discussion of intelligence, Hofstadter describes the difference between a signal and a symbol in the organization of an ant colony (1979, 325); and in its analysis of semantics, continental linguistics describes the "double articulation" of language. Hofstadter's description is, again, probably most accessible in its concreteness. In answer to the question, What does a symbol do that a signal couldn't do? the Anteater replies:

> It is something like the difference between words and letters. Words, which are meaning-carrying entities, are composed of letters, which in themselves carry no meaning. This gives a good idea of the difference between symbols and signals. (325)

The relation between words and letters describes the double articulation of language, the articulation of what André Martinet calls the "semantic content" and the "phonic shape" of language (1962, 26). As we shall see in chapter 4, this "duality of patterning" is, as Michael Corballis argues, "considered a hallmark of true language" and "has also been described as 'one of the most difficult problems of language evolution'" (2003, 114). These descriptions of language are biplanar, noting the separation of the articulation of *meaning* (located in Hofstadter's "words") from the articulation of phonological differences (located in his "letters"). Following Saussure, Martinet argues that the "first articulation" of language constructs a system of distinct signs corresponding to distinct ideas, that it creates its signifying values by means of the systematic differences of its elements. The "second articulation," however, constitutes the minimum meaningful units more simply, solely by means of the presence or absence of physical properties (the distinctive features of phonemes: contact between tongue and teeth, opening and closing of nasal passages, etc.).

Hofstadter's *signal* is the "phonic shape" of a signifying system, subject to the lower level of more or less formal (mathematical) physical descriptions; his *symbol* is the "semantic content" which, as he says (sounding much like Peirce in his description of the interpretant), causes "other symbols to be triggered. When one symbol becomes active, it does not do so in isolation" (1979, 325). Such symbols, he argues, are effected in ant colonies by the "caste distribution" of ants (his analysis follows Wilson's *Insect Societies*). Thus, in the argument, he says:

> Let's lump "caste distribution" and "brain state" under a common heading, and call them just the "state." Now the state can be described on a low level or on a high level. A low-level description of the state of an ant colony would involve painfully specifying the location of each ant, its age and caste, and other similar items. A very detailed description, yielding practically no global insight as to why it is in that state. On the other hand, a description on a high level would

involve specifying which symbols could be triggered by which combinations of other symbols, under what conditions, and so forth. (325)

Here, Hofstadter is describing the "double articulation" of the ants/ant colony in terms of levels. Holland describes a similar double articulation in relation to the "constrained generating procedures" of both computers and emergent phenomena more generally. "Every storage register in the computer," he writes, "contains a string of bits that is interpreted as a binary number. However, when the computer designates that storage register as the next instruction to be executed, that same number is also interpreted as an instruction. Thus duality allows the computer to manipulate its own instructions" (1998, 170). In this description, the number is analogous to the "phonic shape" of language and the "instruction" to its global, semantic content. All of these instances describe the organization or emergence of order out of order: global symbols out of the aggregation of ants in the colony, signaling instruction out of the registrations of memory effected by "persistent patterns" of phenomena in computers and constrained generating procedures, and lexical semantics out of systematic phonemic organization.

The Semiotics of Negation

If in these systems of biological adaptation, artificial intelligence, and the double articulation of linguistics create order out of order (i.e., colonies from ants, persistent patterns that build upon one another, semantic order from phonemic order), what can the "creation" of disorder out of order possibly mean? A. J. Greimas perhaps best articulates this last state of affairs in *Structural Semantics,* one of the most rigorously systematic treatments of language. Greimas argues that the "edifice" of language "appears like a construction without plan or clear aim, like a confusion of floors and landings *[paliers]*: derivatives take charge of classes of roots; syntactic 'functions' transform grammatical cases by making them play roles for which they are not appropriate; entire propositions are reduced and described as if they behaved like simple adverbs" (1983, 133). A clear example is the way a sentence can function as (i.e., play the role of) a noun: "'I like Ike' is my favorite sentence." Greimas summarizes this situation by asserting that "discourse, conceived as a hierarchy of units of communication fitting into one another, contains in itself the negation of that hierarchy by the fact that the units of communication with different dimensions can be at the same time recognized as equivalent" (82).[15] In this description, Greimas offers a detailed—and more or less technical—description of what Bruce Clarke notes as the more general condition of "information": that it "is inherently hyperbolic at all scales; more precisely, it has no proper scale" (2002, 31).

In a similar fashion, though from a thoroughly distinct linguistic tradition, Noam Chomsky makes recursion—the embedding of linguistic phrases within one another that enables a potentially infinitely long statement—a defining feature of language. "A recent review of human language . . . ," Steve Methin argues, "by the equally prestigious Harvard University team of Marc Hauser, Noam Chomsky and Tecunseh Fitch, concluded that recursion is the only attribute of language that lacks any parallel within animal communication systems" (2006, 17). My example of the sentence "I like Ike" functioning as a noun is a special case of linguistic recursion. In chapters 4 and 5, I examine the concept of "theory of mind"—in that, as Daniel Dennett argues, "Even preschool children delight in playing games in which one child *wants* another to *pretend* not to *know* what the first child *wants* the other to *believe* (*fifth-order* intentionality): 'You be the sheriff, and ask me which way the robbers went!'" (2006, 111)—the very articulation of which, as in Dennett's convoluted sentence, is built on recursion (specifically, the recursion of contrary-to-fact constructions).

A key term in Greimas's description is *appropriateness* (when he notes how language can make grammatical cases "play roles for which they are not appropriate"), just as it is a key term when Holland discusses levels of understanding in terms of the appropriateness of frameworks "that are well-tuned to the question(s) at hand" (1998, 185). (Descartes uses the term as well when he notes even the dullest of men can give "an appropriately meaningful answer" [1985, 140].) When Greimas describes the transformation of grammatical cases by syntactic functions to allow them to play roles for which they are not appropriate, he is describing a constant feature of emergent phenomena (most notably in the material homologies of natural selection), one that Nietzsche describes in the *On the Genealogy of Morals* when he notes that "the cause of the origin of a thing and its eventual utility, in actual employment and place in a system of purposes, lie worlds apart; whatever exists, having somehow come into being, is again and again reinterpreted to new ends, taken over, transformed, and redirected by some power superior to it" (1967b,

77; see also Gould's discussion of the "panda principle," 1986, 64–66; 2002, 104; and chapter 4). Holland describes this in the conclusion to his study in his assertion: "*The context in which a persistent emergent pattern is embedded determines its function*" (1998, 226). He offers as a "sophisticated" example the "multifunctionality in biological systems" (an example, I should add, of the material reality of homology). "For instance," he writes,

> a set of three bones that begins by providing a flexible linkage allowing the extra-wide extension of a reptile's jaw and, still later, a linkage in the inner ear of a mammal. The three-bone linkage is well defined and preserved over time, but its function varies according to context. It is this changing aura that makes it hard to characterize emergent phenomena a posteriori. (1998, 226)

Holland's term "aura" is analogous to Hofstadter's "epiphenomenon," and both comport with Nietzsche's "system of purposes."

The problem of purpose is central to any sense of nonreductive materialism, especially purpose inflected toward "purport," Peirce's "law that will govern the future" (1931–35, 1:23). In fact, if I were pressed to define materialism, I might very well say that the best definition of "matter" is phenomena that at once are the sources of sensation, resistence, comprehension (Peirce's categories of signs), alternatively ordered in vocabularies of matter, energy, and "differential" constellations of information without central executives. In any case, the conception of emergence as the emergence of an order that *seems* purposeful even though it does not possess a central "director" of purpose is clear in Holland's example of a computer program for playing checkers that generate "behaviors that transcend the capacity of the designer" (1998, 112) insofar as the computer came to regularly defeat the program designer. Holland's aim, in *Emergence,* like my aim here, is to account for the appearances of purposeful behavior without recourse to transcendental categories of "ghosts" of one sort or another. Holland pursues such an account in his demonstration that the "transformation from the extremely unlikely to the likely is a major characteristic of systems exhibiting emergent phenomena" (231). His argument is remarkable in its ability to offer a "non-positivist" materialist[16] account for natural selection so that, as he says, "what was extremely unlikely on inspection of the generating procedure based on the interaction of atoms to form molecules, becomes likely—almost inevitable—once we take into account the formation of a higher-level generating procedure" (230). Occurrences that seem unlikely often suggest another, external power helping the organization. This surely is what Hofstadter means when he describes the seeming existence of "purposefulness behavior" in an ant colony (1979, 320). Wilson's description of ants as "walking batteries of glands filled with semiotic compounds" (1998, 70) is a reductive description of the phenomena of ant colonies, but while it describes, mechanically, the necessary elements of colony organization—just as the "second articulation" of phonemes and their iconic materiality describe the necessary elements of linguistic organization—it fails to comprehend the "persistent pattern" that makes a colony a colony.

One example of this ability of a system to suggest purpose is apparent in what Chris McManus describes as "the *theory of random cerebral variation*" (2002, 229). This theory examines the "usefulness" of random behavior in complex systems (such as the brain McManus is describing). "Complex systems of any sort," he writes, "show an interesting tension between organisation and randomness. Complexity theorist Stuart Kauffman has shown how the right amount of randomness or chaos facilitates many complex processes" (229). Another, perhaps defining, example of organized randomness is Greimas's description of what I call the creation of disorder out of order in language, an example that offers a way of thinking about the global behavior of complex systems without recourse to ghosts or reductions that fail to see the forest for the trees. At the heart of it—as at the heart of Holland's description of the emergence of nondirected patterns—is the asymmetry between negative and positive phenomena, the very *positive* power of negation. That is, on the level of mechanistic physics—the level of the *aggregation* of phenomena—negation itself does not exist: phenomena are either present or absent. And absence—of an oxygen atom in water as opposed to its presence in hydrogen peroxide, for instance—is not a determination of the phenomena at hand: the relationship between water and hydrogen peroxide is in no way hierarchical; water and hydrogen peroxide are not different classes of fact. On this level, presence and absence are symmetrical.[17] On the level of biological and environmental sciences—that is, on the level of the *interaction* of ordered systems—different classes of fact are readily discernible in the very *asymmetry* of nested levels I have been describing, the *constraining* power of the level of aggregative physics upon the interactions of biological systems.[18]

The very *adaptiveness* of biological systems and, in Holland's discussion, emergent phenomena altogether (including the "emergences" of the auras and epiphenomena of "mind" and "intelligence"), is a function of the pragmatic *difference in class*. This can be seen in Holland's use of the terms *good* and *bad* outcomes. Good means the continued persistence of complex patterns, capable of future elaboration and further complexity; bad means extinction, the end of the game, with no further revision or elaboration possible. The very idea of hierarchy is based on the constraining power of lower levels on higher levels (or, to put it positively, the *presupposition* of the lower by the higher). This is why Holland argues that the checkers program "should concentrate on avoiding bad outcomes, rather than directly seek good outcomes. Bad outcomes should be avoided whatever the cause, while good outcomes may not hold up against other [more competent, more powerful] opponents" (1998, 71).

On the level of semiotic negation, the difference of class (Peirce's different kinds of fact) is not absent, as in mechanical materialism, nor is it functional, as in systematic adaptation. Rather, it becomes a problem: the problem I articulated in my awkward phrase, "positive negation." Brian Rotman has described what I call the positive power of negation as the "systematic ambiguity" (1987, 2) inherent in the sign for zero, though Heisenberg's term, "undecidability," might be better than ambiguity. Zero, Rotman argues, has a "double aspect . . . , as a sign inside the number system and as a meta-sign, a sign-about-signs outside it [a sign, that is, signifying the absence of numerical signs], that has allowed zero to serve as the site of an ambiguity between an empty character (whose covert mysterious quality survives in the connection between 'ciphers' and secret codes), and a character for emptiness, a symbol that signifies nothing" (13). Such ambiguity, Rotman takes pains to explain, should "be seen not as an error, a confusion to be clarified, but as the inevitable result of a systematic linguistic process" (3) which allows for levels of organization. Negation articulates both a *contrary* relationship, A versus –A, between the extreme poles on a continuum, such as white versus black on the continuum of colors or zero versus infinity on the continuum of numbers; and it also articulates a *contradictory* relationship, A versus ~~A~~, the presence and absence of a category—what Kaplan calls its "invisibility" (the invisibility of colors or of heat)—on a *lower* level of analysis, such as colorlessness (e.g., black as the absence of light) versus colorfulness (e.g., white as all colors, as light energy). Black is both sign and metasign: one color on a many-colored pallet which, like zero on the pallet of numbers, can be combined in many ways with many different colors. But on a lower level of analysis, on which light and light energies are the measures, talk about colors is meaningless.

This description follows A. J. Greimas's discussion of the "semiotic square." (For a detailed discussion, see Schleifer 1987, xx–xxiii, 25–33.) The crucial issue, I think, is the presence versus absence of categories, because here the logic of different classes of fact is articulated. This is most clearly discerned in inscribing Hofstadter's discussion of the vocabularies of heat on a semiotic square.

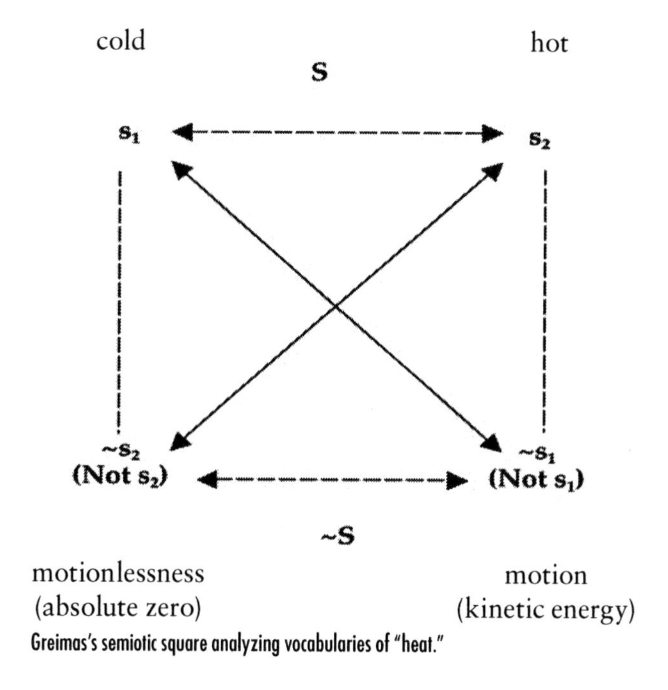

Greimas's semiotic square analyzing vocabularies of "heat."

"Motionlessness" is the absence of "hot" when hot is reconceived as another class of fact, kinetic energy rather than temperature; and "cold" is comprehended as another class of fact from kinetic energy when it is conceived as a relative temperature rather than a measure of motion.

Let me say this more clearly: negation is the result of comprehended wholes, and it results from and results in the disorder of parts and whole. (This might help define Peirce's elusive term "symbol" as well.) This is what Rotman describes as the *systematic ambiguity* of semiotics analysis altogether. That is, if mechanical physics—

positivist science—deals with parts in which the whole is the sum of the parts, and if the interactive biologies—adaptive science—deal with wholes which are greater than the sum of the parts (see Capra 1997, 17; Prigogine and Stengers 1984, 143–44), then semiotics, what I call negative science, deals with systematic ambiguity—or better, *systematic undecidability*—in which the "part" is alternatively simply an accident and thus not a part at all; in which, that is, a so-called part is alternatively external as well as internal to the whole. Black is external to the light spectrum, its contradictory, though it also finds itself on the spectrum, a color among colors; cold is external to measurements of heat, though it also finds itself a "kind" of temperature. Like Greimas's example of discourse that alternatively recognizes differences and equivalence in "units of communication with different dimensions [that] can be at the same time recognized as equivalent" (1983, 82), the example of zero—the defining example of the negative science of semiotics—presents a phenomenon that is both part of (internal to) the system of numbers and contiguous with (external to) that system.[19] Finally, we can see in the semiotics of negativity the possibility of the concept of "level" itself. Lower levels, as Holland notes, "constrain" upper levels and to that extent are, in fact, "parts" of them. But such constraint behaves negatively insofar as the terms and vocabularies of different levels—the very *parts* that comprise a level—do not "translate" across levels. "Zero," Rotman argues, "functions dually: it moves between its internal role as a number among numbers and its external role as a meta-sign initiating the activity of the counting subject" (1987, 31).

The tripartite description of levels I pursue conforms in interesting ways to Peirce's tripartite description of signifying practices under the categories of icon-index-symbol. The icon offers correspondences part for part (or at least experience for experience); the index offers *motivated* correspondences, symptom for disease; and the symbol offers unmotivated correspondences, what Saussure describes as the arbitrary nature of the sign. Similarly, mechanical positivism is the science of elemental parts; biological adaptationalism is the science of seeming purposefulness; and the systematic undecidability of semiotics is the science of the confusions of parts and accidents.

Homologous Levels

As I suggested in the introduction, a homologous correspondence in structure but not function governs the relationships among the levels of understanding I am examining. Let me explain. As David Chalmers suggests, a materialist (or physicalist) explanation focuses solely on structure and function.[20] Chalmers agrees that phenomenal experience has structure as well as experiential content and that both structure and content are "grasped together." However, Chalmers contends that the "usual materialist explanations" of things are explanations solely in terms of structures and functions (and dynamics)—and structure and function cannot alone, he argues, also explain or account for the content of experience. Thus, the usual kind of materialist explanation has no place in it for "experiential content," because in the usual sort of materialist explanation, the combination of structure and function does all the explaining and is all that is explained. Structure and function alone can only explain more (and more complex) structure and function. So such explanations cannot (on their own) also explain content. That's not a problem for explanations of most kinds of physical/material things, since for them we need no account of content—we only need an account or explanation of form and function. Thus, we describe how light works in terms of the structure of electromagnetic waves and how such waves interact with other material things because of their structure and the structure of those other material things (see Casti 1995, 65–69 for an analysis of light caustic in relation to catastrophe theory). But in the case of phenomenal experience, the "phenomenal content" also needs to be accounted for (or explained). And, Chalmers argues, no explanation in terms only of structure and function can do this.

On the level of biological systems, and especially natural selection, explanation homologously organizes itself in terms of structure and function. Natural selection provides an explanation of how species have come to be (how they have evolved) and of the particular traits that predominate among members of those species. The explanation is in terms of the presence of variations in heritable traits among members of ancestral breeding populations and how in a particular environment certain of those differing heritable traits lead to a reproductive advantage (i.e., higher level of fitness) for their bearers, which tend to result in their producing more offspring (who in turn carry those favorable traits). At the level of organisms within a species, the sorts of structures and functions explained are bodily structures and behavior patterns. The functions explained are the roles (functions) these have played in the survival and reproductive success of their ancestors. And, at a "higher level" of organization, the structures explained may be the structures of a whole ecosystem (i.e., the diversity of species, the traits that predominate among members of those species, and how those species interact) at a given time in evolutionary history. The sorts of functions of the ecosystem explained include the "apparent balance of nature" (between predators and prey, for example). Chalmers takes it for granted that explanations of this sort, explanations of structures and functions in organisms and in ecosystems, are ubiquitous in evolutionary biology—they are largely what evolutionary biologists spend much of their time working on (for specific species or specific environments).

Chalmers takes these to be material/physical explanations of just the kind he is talking about, yet as Gould notes, these explanations do not "prove" themselves by means of prediction. Rather, they are retrospective accountings of phenomena. When McManus claims that a "theory of random cerebral variation"—an "unusual theory," he says, that "celebrates variation and difference, and provides an explanation for the lay belief that some people literally 'think differently' or have their brain 'wired differently'"—he notes that "the heart of Darwin's theory of natural selection is that mutations occur at random and that despite many being disadvantageous, selection picks out the few that are beneficial and maintains them" (2002, 229). He notes that "complex systems of any sort show an interesting tension between organisation and randomness" and that "life itself occurs on the edge of chaos, tiny alternations precipitating larger changes that are maintained without being destroyed" (229).[21] It is the "edge of chaos," he claims—where there is randomness but "not too much randomness, for otherwise it would destroy the very thing that it tries to create" (229)—that conditions the creation of order out of order Schrödinger discusses. In a hypernote (http://www.righthandlefthand.com) McManus adds, "The same principles of organisation and randomness can be seen in large social organisations, particularly where a slight rearrangement of an otherwise well-functioning system can sometimes benefit the overall workings of the system." With this note, he includes in his discussion, as I do here, levels of mathematical formalism, biological explanation, and social organization.

These modes of explanation have the same structure, even while their goals—*prediction* and what Gould calls "postdiction" (1986, 65)—are different. The goal—or at least one goal—of semiotics (and even information theory) is what I call *speculation*. Earlier, I mentioned that MacKay argued that the reception of the message was as important as its emission, that, as Greimas says in an altogether different context, linguistics was the science of perception, not expression. Such a "perceptible" science necessarily traffics in semantics and "meaning"—which is to say *content*—but as a science it attempts to understand meaning (as all sciences do) in terms of structure and function. Greimas claims that insofar as language is "conceived as a hierarchy of units of communication fitting into one another," it participates in structure and function, even though it "contains in itself the negation of that hierarchy by the fact that the units of communication with different dimensions can be at the same time recognized as equivalent" (1983, 82). In this way, just as the focus on random occurrences in natural selection disrupts "system"—or at least asks us to reconceive "system" outside of always repeatable phenomena—so semiotics disrupts "function," since the very communicative functioning of language is disrupted in the breakdown of hierarchy. Moreover, both natural selection and semiotics do so, in their different ways, by taking history and narrative, respectively, into their purview.

This is an implication, I think, of Gould's insistence that *external* factors such as a meteor, as well as random mutation that McManus mentions—which is based on the (more or less systematic) instability of replication—has, in fact, affected the history of natural selection (for an extended account of external factors, see Wiener 1995). On a different level, Greimas's point is that linguistic elements can organize themselves in "grammatical" structures beyond the sentence that makes narrative possible. Gould's focus on history and Greimas's focus on meaning make the "content" that Chalmers argues is not part of materialist/physicalist explanation the focus of attention. But making these different contents the objects of explanation does not abandon the *form* of explanation—namely, an account of structure and function—but it transforms, homologously, the relation between those elemental forms and the goals of their combination. Thus, in mathematical physics, evolution, and semiotics, the organization of structure/function is taken up to different ends. In the terms I developed from Peirce in chapter 1, there is the "logic" of *content-free* explanation of physicalist explanation, the "induction" of *contingent historical contents* in evolutionary explanation, and the "abduction" of *contrary-to-fact narrative content* in semiotic explanation. The test of physicalist explanation is *prediction;* that of evolution, as Gould says, is *postdiction;* that of semiotics *speculation.*

The homological connections among these levels might become clear in a strictly mathematical example, since mathematics in important ways is the purest of "content-free" explanations. (In fact, Bertrand Russell has asserted that "mathematics may be defined as the subject matter in which we never know what we are talking about, nor whether what we are saying is true" [1917, 71]). John Casti, in examining catastrophe theory—"a mathematical framework" for describing "discontinuous processes" in which "small changes in the input . . . can lead to a big change in the final result" (1995, 45)[22]—also surveys these levels of understanding in examples of the work of catastrophe theory in physics, developmental biology, and cultural history. Specifically, he demonstrates the uses of the mathematics (geometry) of catastrophe theory in relation to the reflection of light in a coffee cup (light caustic) in a physical phenomenon, the differentiation of cells (morphogenesis) in embryology, and the phenomenon of the collapse of civilizations in history. He notes that when "we know the precise mathematical form of the family members," as we sometimes do, "especially when the system under investigation is one from physics," then "we can say a lot more about the situation, perhaps even being able to make some quantitative predictions about what the system will do" (55). He also notes that René Thom "originally developed catastrophe theory as a mathematical way of addressing . . . the problem of cellular differentiation and morphogenesis, or the emergence of form" (69). Yet, Casti argues, "developmental biology is an area somewhere between physics and philosophy when it comes to being able to write down an explicit mathematical representation for the underlying dynamical processes. We know *something* about morphological processes. But we don't know nearly as much as we know about phenomena like the light caustic studied earlier" (73). And when we turn to factors contributing to "the collapse of early state societies," he says, "we have clearly moved from the realm of the physical to the metaphysical insofar as our use of catastrophe theory goes" (77). In this gathering of examples, Casti describes "the transition from physics to 'semi-physics,'"—"in the biological sciences, we must take refuge in a kind of semiphysical way" (78, 79)—and then to "social and behavioral arenas" in which "the basic approach is to use what has come to be termed the *metaphysical way* of catastrophe theory" that is governed by "postulation," "assumption," and "presupposition"—terms that represent "out-of-the-blue assumption[s] that we must be ready to swallow if we want to appeal to [catastrophe theory] to single out a particular geometry describing our problem" (78, 79–80).

The difference between physics and biological sciences, he argues earlier (in a discussion congruent with Gould's description of "historical" sciences) is the difference between prediction and explanation. "We laid great emphasis," he says,

> upon the fact that the scientific answer to a question takes the form of a set of rules, rules having specific properties and modes of generation. Usually these rules are encoded in the form of a mathematical model or, more generally these days, a computer program. Moreover, the rules are used in two quite distinct ways: (1) to *predict* the outcome of future observations and/or (2) to *explain* past observations. Newton's laws of celestial motion are the quintessential example of a set of rules used for the first task, while the principle of natural selection in evolutionary biology exemplifies the second. Of course, in some cases the same set of rules can be used for both purposes, as with Newton's laws of motion. But this is generally not the case. (1995, 47)

In this opposition, Casti seems to be elaborating Schrödinger's distinction between physics and biology. Whether or not this is technically correct, however, this distinction can help me delineate the homologies I am pursuing. If prediction and explanation are the goals or purposes of the application of the complications of catastrophe to physics and biology respectively, then *speculation*—"out-of-the-blue" assumptions—is the goal or purpose of the human or semiotic sciences.[23] Such goals nevertheless utilize the same mathematics, taking up for different ends (as history shows us the "material reality" of homological development does) the same formal procedures.

Error and Confusion

I should like to substantiate this typology allowing for an intangible materialism by examining evidence that Wilson sets forth in *Consilience.* To this end, I to return to the Cartesian story and its contrary comprehensions of the doll's appearance, especially of her—or is it "its"?—face. The recognition of human (and especially female) faces is a significant theme in Wilson's book. It is equally a theme in the philosophy of Emmanuel Levinas, a memorable figure in Wittgenstein, a narrative modality in Elizabeth Bowen's late novel *The Heat of the*

Day, and, neuroscience informs us, physically inscribed in "face-responsive [neuronal] cells" of the human—and primate—brain (Aggleton and Young 2000, 113; see also Emery and Amaral 2000, 179). An examination of Wilson's treatment of faces can well underline both the strengths and the weaknesses of his enlightened reductionism. Faces first appear in *Consilience* when Wilson is talking about the structure of the brain, including a "final step [in which] the brain was catapulted to a radically new level, equipped for language and culture" (1998, 106). Cataloging neurological research, he describes one "bizarre result" of disturbances of particular brain circuits, namely prosopagnosia, in which "the patient can no longer recognize other persons by their faces, but he can still remember them by their voices" (108). In this discussion, he examines the physiology of the brain conceived of as a machine. The next discussion of faces takes place in the context of Wilson's analysis of the "primary epigenetic rules" imposed by the senses, the manner in which they break "otherwise continuous sensations into discrete units," such as classifying the differing wavelengths of light into four basic colors and automatically dividing "continuous speech sounds into phonemes" (151). The face, Wilson argues, "is the chief arena of visual nonlinguistic communication and the secondary epigenetic rules that bias their psychological development," and he goes on to focus on the mouth and smiles (152–53). Wilson's third examination of faces—his longest—focuses on "the beauty of a young woman's face" (230). Although these three examples seem to emphasize physiological cause and effect, the organism's interaction with its environment, and the semiotics of beauty, in fact in each instance Wilson strives to underscore the continuity of these levels of explanation, the interchangeability among the mechanical monstrosity of bizarre results, the adaptive interactions of smiles, and the meanings generated by "epigenetic rules relevant to aesthetics" (230).

His last example of female human attractiveness is a problem for natural selection conceived as a mechanical process and, by extension, for the mechanical conception of meaning he presents earlier when he argues that the properties of language and culture are selected, mechanically, for reproductive success. ("What we call *meaning,*" he asserts, "is the linkage among the neural networks created by the spreading excitation that enlarges imagery and engages emotion" [115].) The problem is this: if the female facial features Wilson catalogs as universally "thought most attractive"—"relatively high cheek bones, a thin jaw, large eyes relative to the size of the face, and a slightly shorter rather than longer distance between mouth and chin and between nose and chin" (230; see also Mithen 2006, 189, for the attractiveness of symmetry)—if these features are the most attractive, then why is it, he asks, that "few women—extremely few in fact—approach" this optimum (230). In other words—and these are *his* words—following his assumption that *all* explanation is reducible to the model of demonstrating the creation of order out of disorder, he asks: "If the perception of facial beauty resulted in the higher survival and reproductive success of the most beautiful conceivable, then the most beautiful should be at or close to the average within the population" (230–31).

He answers by switching levels of explanation:[24] he describes what he calls "the behavioral phenomenon known as the supernormal stimulus" often found in animal species, which he asserts is "the preference during communication for signals that exaggerate the norms even if they rarely if ever occur in nature," a preference that is "widespread among animal species" (231). His example is the silver-washed fritillary, an orange butterfly. The males of the species have been found to "really prefer" simulated females, plastic replicas whose wings are "the biggest, brightest, and most rapidly moving" (231) even though no such females exist or—given the size of these plastic replicas—could possibly exist in the environment. "In the real world," Wilson argues, "the supernormal response works because the monstrous forms created by experimenters do not exist" (231). "In a parallel manner," he argues, the ideal of female beauty—presenting "physical cues of youth, virginity, and the prospect of a long reproductive period" (231)—may well have adaptive significance. Leaving out the virginity, which we may well argue over, the switch from nonexistent and nonadaptive huge butterflies to the "extremely few" existent human beauties is a sleight of hand. Moreover, the very importation of "preferences during communication" as *something to be measured* underscores a shift in the levels of explanation. The levels scattered throughout *Consilience*—to quote Wilson, "events of physical causation" (11), the adaptation of "organisms to their environment" (128), and intraspecies forms of communication (including the "walking batteries of glands filled with semiotic compounds" [70] Wilson sees in his ants)—stage themselves, so to speak, in resting places and nests upon which other modes of explanation are built.

We can see this in the many faces of consilience. Let's begin, as

Wilson does, with "reductionism, the breaking apart of nature into its natural constituents" (54). The natural constituents of face he describes are mouth, cheek, jaw, eyes, chin, nose, as well as eyebrows and pupils he mentions elsewhere. In his discussion, the disorder of parts turns into the order of physiological causation. That order expands to the larger order of adaptive interaction. And those adaptations return—across different cultures—to the arbitrary nature of the sign. "Throughout life," he writes, "smiling is used primarily to signal friendliness and approval. . . . [But] each culture molds its meaning into nuances determined by the exact form and the context in which it is displayed" so that "smiling can be turned into irony and light mockery, or to conceal embarrassment" (153). The hard wiring of the molecular biology of the brain—both primary and secondary epigenetic dispositions and predispositions—creates order out of disorder in the service of the biological creation of order out of order Schrödinger discusses, the fact that biological systems are built out of other biological systems. I want to describe the third level of semiotics not only in the aesthetics of beauty Wilson mentions or in the waving good-bye Shapin mentions, but in Wilson's own remarkable sentence: "In the real world the supernormal response works because the monstrous forms created by experimenters do not exist, and the animals can safely follow an epigenetic rule expressible as follows: 'Take the largest (or brightest or most conspicuously moving) individual you can find'" (231). In this sentence, Wilson literally *negates* cause: he proffers a negative cause—"it works *because* the monstrous forms do not exist"—just as the mechanical monster, Francine, both exists and does not exist (being no monster but simply a Cartesian/Wilsonian organism-machine, and besides, the wench is dead) inside and outside the "level" of meaning.

Two more faces can complete this picture of the work of semiotics, creating, as I argue, disorder out of order. Here is Emmanuel Levinas talking about the meaning of face:

> The proximity of the other is the face's meaning, and it means from the very start in a way that goes beyond those plastic forms which forever try to cover the face like a mask of their presence to perception. But always the face shows through these forms. Prior to any particular expression and beneath all particular expressions, which cover over and protect with an immediately adopted face or countenance, there is the nakedness and destitution of the expression as such, that

> is to say extreme exposure, defencelessness, vulnerability itself. This extreme exposure—prior to any human aim—is like a shot "at point blank range." . . . From the beginning there is a face to face steadfast in its exposure to invisible death, to a mysterious forsakenness. (1989, 82–83)

In this passage, Levinas answers the powerful optimism—the Enlightenment optimism—of Wilson's mechanical materialism with the darker meaning of that materialism. Here is another materialist consilience in an ethical understanding—Wilson ends his book with a call to physico-biological ethics—that, for Levinas, is not a superstructure to materialism's base but rather itself a base beyond the plastic forms of positive materiality. Like the sailors on Descartes's ship, Levinas apprehends the death inscribed in mechanical materialism as a meaning—of defenselessness, vulnerability, forsakenness—that names the disorder of the physical world and precisely calls for the ordered interactions with that world as a species of responsibility prior to and at the base of mechanical response. However adaptive face-to-face encounters are, they are first of all—before all practical use—inscribed with "a summons of responsibility": "in the face of the other man," Levinas writes, "I am inescapably responsible and consequently the unique and chosen one. By this freedom, humanity in me *(moi)*—that is humanity as me—signifies, in spite of its ontological contingence of finitude and mortality, the anteriority and uniqueness of the non-*interchangeable*" (84). Before interaction, before the mechanics of finitude and mortality, is the disorder of noninterchangeable meaning. Such noninterchangeable meaning creates—or manifests itself as—what I call the power of language.

If Levinas is negative in theme—capturing in face the moment of disorder in the physical world's orderliness and gathering up in face, prior to the eyes and mouths and noses of particular faces, an apprehension of responsibility that allows face to appear strange and familiar at the same time—then Elizabeth Bowen offers a negative discourse of facial expression. Near the beginning of *The Heat of the Day*, she describes a person who we later learn is a relatively minor character in the novel.

> He confronted a woman of about twenty-seven, with the roughened hair and still slightly upward expression of someone who has been lying flat on the grass. Her full, just not protuberant eyes looked pale in a face roughly burned by summer; into them the top light of the

> roofless theatre struck. Forehead, nose, cheekbones, added no more than width. Her mouth was the only other feature not to dismiss; it was big; it was caked round the edges, the edges only, with what was left of lipstick inside which clumsy falsified outline the lips turned outwards, exposed themselves—full, intimate, woundably thin-skinned, tenderly brown-pink as the underside of a new mushroom and like the eyes once more, of a paleness in her sun-coarsened face. It was the lips which struck him and could have moved him, only that they did not. Halted and voluble, this could but be a mouth that blurted rather than spoke, a mouth incontinent and at the same time artless. (1962, 11)

Here is a face that offers not the forlornness of a physical world without spirit or the call to a kind of species-ethics in the face of that disenchanted world but simply a discourse of negation that proceeds, as Wilson might say, *because* of things that do not exist: eyes that are just not protuberant, features that add nothing but the Cartesian primary quality of extension, a mouth that is not to be dismissed, lips that could have moved him but do not, colorless paleness, and blurted speech beyond the borders of continent reason. If Wilson's mechanical physics is a positive science, this is the negative science of contrary-to-fact. The defining feature of any semiotic system, Umberto Eco has argued, is its ability to tell a lie (1976, 7); more generously, George Steiner has argued that counterfactual conditionals and the future tense—precisely because they do not exist in the real world—comprise the genius of language in which, he says, "we can *say anything*" (1975, 216).

Chief among those sayings are the possible worlds we construct, consilient or forsaken; the promises we make to one another, making tools, building communities and intellectual disciplines; and the levels we perceive in our understanding and the meanings and feelings we apprehend in one another's faces. Wilson argues in *Consilience* that "the capacity for remembrance by the manipulation of symbols is a transcendent achievement for an organic machine" (1998, 111). He also argues that "contract formation is more than a cultural universal. It is a human trait as characteristic of our species as language and abstract thought, having been constructed from both instinct and high intelligence" (171). A bit earlier he says that "*contractual agreement* so thoroughly pervades human social behavior, virtually like the air we breathe, that it attracts no special notice—until it goes bad" (171). He does not seem to see that the "going bad," like the aesthetics of beauty that seems to break the mold of natural selection, breaks the system of positive science. "More than error," Wilson writes, "more than good deeds, and more even than the margin of profit, the possibility of cheating by others attracts attention. It excites emotion and serves as the principal source of hostile gossip and moralistic aggression by which the integrity of the political economy is maintained" (172). If, in fact, contract formation is adaptive, why is it that it goes bad? What demon inhabits the basic laws of positive physics to allow the rule of law to flounder? Surely Wilson's ants and Descartes's mechanical doll never break the law of their makeup, never put their lips to other uses than smiling, their eyes to other effects than the stimulations of light waves, their faces to anything beyond facing the world. The semiotic compounds of ants exist on a different level from the semiotic compoundings of our meanings; the laws of their functioning are different.

At the beginning of *Consilience,* Wilson describes "three preconditions, three strokes of luck in the evolutionary arena" that led to the scientific revolution. All three, I suggest, point to levels of understanding beyond the base of positive physical attributes. The first, Wilson says rather vaguely, is "the boundless curiosity and creative drive of the best minds." Another enabling precondition he describes is "what the physicist Eugene Wigner once called the unreasonable effectiveness of mathematics in the natural sciences." And a third, he says, is "the inborn power to abstract the essential qualities of the universe." "This ability," he adds, "was possessed by our Neolithic ancestors, but (again, here the primary puzzle) seemingly developed beyond their survival needs" (48). All three of these preconditions are different versions of understanding apprehended on the levels of the "creative drive" of semiotics, of the mechanical–physical world described by the mathematical sciences, and of biological interactions with the environment. The fact that they each developed beyond the boundaries of perception; beyond parsimonious, necessary, and sufficient reason; and beyond our survival needs is not puzzling if they are understood as nested stages of understanding, stretching from disorder to order and the orderly disorder of meaning—from the mechanical physics of mouth to adaptive smiles and ironies of what mouths do not quite say. I suppose there's a consilience here as well, but one that cannot, I'm afraid, lend itself to the parsimonies of "physical causation" (11), the "one class of explanation" Wilson describes, the "seamless web of cause and effect." On top of that single class exists two others: the means and ends of natural selection, order out of order, and the disordered order of semiotics in which positive and negative facts, promises and fraud, hierarchy and its opposite, jostle together beyond the necessities of cause and effect and the functionalism of means and ends to create our meanings.

The next three chapters examine the levels I have described here in relation to material human attributes, voice, hands, and pain. These examinations focus on the physical basis of Tourette syndrome in the neurochemical functioning of the brain, the human hand as the product of evolutionary development of order out of order, and the ways that the ubiquity of pain lends itself to negative semiotics, most dramatically instanced in the phenomenon of phantom pain. But each chapter describes all three levels as well: the ways that the dysfunction of neurological systems in Tourette syndrome suggest the adaptiveness of passionate language; the physiological basis of the evolution of language, hand, and pain; and the ways in which the materiality of these bodily phenomena create contrary-to-fact situations that can allow us to discern the disorderly lawfulness that will govern the future. In these chapters, I demonstrate the functionality of nestings for our understanding of "fact" and process and, at the same time, demonstrate a sense of materialism that transcends and encompasses these nested levels. I also suggest the material basis of literature and Peirce's symbolic altogether: the material basis of the sensuous affectiveness of language, of the material situation of the speaking subject in the world where things are indicated and organized in narrative forms, and of the intangible materiality of subjectivity—including spiritual and religious subjectivity—measured against, and sometimes in terms of, pain. Out of these discussions, I hope the delineations of the power of language, and especially literary language—its material sensuousness, its ability to attend to and on the material world in reference and narrative organization, and its taking up of material experience in order to shape a sense of the subject of experience—might emerge.

3. Material Voices: Tourette Syndrome, Neurobiology, and the Affect of Poetry

The neural bases of human language are intertwined with other aspects of cognition, motor control, and emotion.

—*Philip Lieberman,* Human Language and Our Reptilian Brain

The "I" of the lyricist . . . sounds from the depth of his being: its "subjectivity" in the sense of modern aestheticians is a fiction.

—*Friedrich Nietzsche,* The Birth of Tragedy

In this chapter, I examine the relationship of poetry to the neurobiological condition known as Tourette syndrome in order to describe a kind of physiological materialism that situates itself within the hierarchy of levels of organization I examined in the preceding chapter. Tourette syndrome is clearly an organic condition that involves, among other symptoms, the seeming emotion-charged use of language, the spouting forth of obscene language that, as researchers note, "may represent," among other symptoms, "a common clinical expression of underlying central nervous system dysfunction" (Fahn and Erenberg 1988, 51).[1] That Tourette syndrome entails the automatic outpouring of emotionally charged uses of language allows us to situate it within an evolutionary framework. In *The Descent of Man,* Darwin notes the ubiquity of music and impassioned speech in all known cultures and argues that "these facts with respect to music and impassioned speech become intelligible to a certain extent, if we may assume that musical tones and rhythm were used by our half human ancestors, during the season of courtship" (cited in Mithen 2006, 178). That "impassioned speech" was selected for its adaptive capabilities suggests that the dysfunction of such speech in Tourette syndrome could be understood within a biological–evolutionary framework as well as the biophysiological frame I foreground in this chapter. The language of Tourette syndrome—with its rhymes, echoes, and strong rhythms—seems to me to be the very kind of "vocal gestures" that Steve Mithen traces in the evolution of language through music. Finally, the fact that poetry takes up many of the resources of the impassioned phonic or linguistic behavior of Tourette syndrome originating in the subcortical—or what many call the "reptilian"—brain suggests a third possible level of understanding the material phenomena of Tourette syndrome.

The uncanny verbalizations of Tourette syndrome, as David Morris (among many others) has argued, are apparently connected "to subcortical structures [of the brain] that permit them to tumble out unbidden, like a shout or cry" (1998, 170). Poetry also, in the description of A. J. Greimas, attempts to create the "meaning-effect" of a "primal cry," an "illusory signification of a 'deep meaning,' hidden and inherent in the plane of expression," in the very sounds of language (1970, 279; my translation). Language, as neuroanatomy has demonstrated, involves various regions of the brain, especially the Broca area in the frontal region of the neocortex and the Wernicke area in the posterior area of the cortex. Both the cortex and neocortex have been consistently associated with more abstract modes of reasoning.[2] But subcortical regions of the brain, especially the thalamus, the hypothalamus, the amygdala, and the basal ganglia—regions that have been called our reptilian brain because humans and other primates inherit them from earlier and less complex life forms—have also been associated with language (see Lieberman 2000). Studies in experimental neurobiology have closely correlated these areas of the reptilian or "old brain" with motor activity, basic instincts, and emotions.

Poetry, I contend, in its more or less intentional and willful activity, calls upon all of these neurological resources of language—so that in poetry, as in the neurobiology of language more generally, the strict distinction between language and motor activities is less and less apposite. This contention, I believe, is illuminated by an examination of Tourette syndrome in its more or less unintentional and impulsive activity. Just as the facial tattoos of Maori warriors create the "effect" of the facial signaling of aggression (McNeill 1998, 302), which is part of the behaviors of many primate species and has clearly been associated with cortical and subcortical regions of the brain (especially the amygdala, the seat of emotions in primates containing what researchers describe as "face-responsive [neuronal] cells" [Aggleton and Young 2000, 113; see also Emery and Amaral 2000, 179]), so poetry creates the effect of the vocal signalings of primates, which, it seems clear, manifest themselves involuntarily in Tourette syndrome. In this chapter, I argue that the conventions and resources of poetry and of what Roman Jakobson calls the "poetic function" of literary language more generally—fascinations with the sounds and rhythms of language, with rhymes and repetitions, with its chants and interpersonal powers—haunt the terrible and involuntary utterances of Tourette syndrome in its powerful connections between motor activity and phonic activity.

By "poetic function" Jakobson is describing that aspect of language that exists outside its referential, emotive, conative, communicative, and semiotic functions. He most fully describes this in "Linguistics and Poetics," where he argues that "any attempt to reduce the sphere of the poetic function to poetry or to confine poetry to the poetic function would be a delusive oversimplification" (1987, 69). "Poetics in the wider sense of the word," he writes, "deals with the poetic function not only in poetry, where this function is superimposed upon the other functions of language, but also outside poetry, where some other function is superimposed upon the poetic function" (73). Derek Attridge situates Jakobson's poetics within the opposition between conceiving poetry as functioning "to heighten attention to the meanings of words and sentences" and conceiving poetry as "a linguistic practice that specifically emphasizes the material properties of language . . . [that] provide pleasure and significance independently of cognitive content" (1988, 130). The second of these conceptions, that of the "poetics" of the material properties of language, can be discerned in the neurobiological dysfunctions of Tourette syndrome. In fact, many of the technical descriptions Jakobson offers as manifestations of the poetic function—including paronomasia, echo rhyme, alliteration (1987, 70), and, more globally, focus on syllabic phonemes (73), which is to say the material sounds of language—describe phonic symptoms of Tourette syndrome. The second of these conceptions emphasizes the reception or apprehension of poetic discourse rather than its intentional production; it emphasizes the *iconic* aspect of signs that Peirce describes. The first of these functions, that of heightening attention to meanings and the situation of discourse, emphasizes the ways that poetry organizes experience; it emphasizes the *indexical* aspect of signs that I explore in the following chapter.

But before I begin in earnest, let me make clear what I am *not* doing. I do not want to suggest that Tourette syndrome is not a terrible ailment, occasioning powerful distress and appalling disruptions in people's lives. Oliver Sacks makes this clear in his book *Awakenings*, in which he describes the "immense variety of involuntary and compulsive movements [that] were seen" in postencephalitic patients after they were treated with L-DOPA, and that included virtually all of the involuntary and compulsive symptoms of Tourette syndrome I describe in this chapter.[3] Describing these symptoms shared by his patients and people suffering from Tourette syndrome, Sacks quotes a line from Tom Gunn's poem "The Sense of Movement": "One is always nearer by not being still." "This poem," Sacks writes, "deals with the basic *urge* to *move,* a movement which is always, mysteriously *towards.*" This is not so, he says, for the patients he encounters: they are "*no* nearer for not being still. [They are] no nearer to anything by virtue of [their] motion; and in this sense, [their] motion is not genuine movement" (1999, 16–17).[4] In the same way, the motor/phonic symptoms of Tourette syndrome I describe do *not* constitute poetry: the language uttered by people who have Tourette syndrome may no more resemble poetry than their involuntary movements resemble pantomime (see Sacks 2007, 227). But the powerful connections between linguistic and motor activity—between the meanings and materialities of discourse exhibited in the meaningless rhymes, rhythms, and invectives of Tourette syndrome—manifest, I believe, the fact that the sources and resources of poetry are seated deeply within our primate brains.

The Nature of Tourette Syndrome

Let me begin with some definitions. Here is a neurologist, Dr. James Miller, describing Tourette syndrome:

> Tourette syndrome is an unusual chronic medical condition characterized by the childhood onset of multifocal motor and vocal tics, which may include phonation, vocalization, or the articulation of

formed words, including obscenities or profanities. Some of the tics relieve an inner urge or compulsion. Persons with Tourette syndrome frequently also have obsessive-compulsive disorder and attention deficit disorder. . . .

A tic is difficult accurately to describe. It is a relatively brief, episodic, muscle contraction, whose frequency and severity waxes and wanes, and which can be mimicked. . . .

> A *simple motor tic* is a sudden, brief, contraction of a localized muscle group, such as to cause the face to twitch or shoulder to jerk.
> A *complex motor tic* involves many groups of muscles and mimics a gesture inappropriate to the circumstance [such as] . . . facial grimacing, touching, or hand wringing.
> A *simple phonic tic* is a noise without meaning such as a grunt, bark, or squeak.
> A *complex phonic tic* is the articulation of meaningful words and includes coprolalia [saying obscene words], echolalia [repeating other people's words], and palilalia [repeating one's own words].
> A *compulsive tic* is a response to an inner urge and may manifest as any of the above. (2001, 520)

Miller offers a description of Tourette syndrome from the vantage of mechanistic physiology; even when the etiology is not known, symptoms are described in a discourse that anticipates a mechanical description based on the search for the elemental "basis" of the physiological condition.

A second definition describes Tourette syndrome by means of the history of this disorder. In *A Cursing Brain? The Histories of Tourette Syndrome,* Howard Kushner offers what might be called an operational definition of the condition, a definition that begins by focusing on how it works rather than focusing on its constituent parts (see Robinson 1972). One narrative he gives is a person with Tourette syndrome blurting out as he makes a airplane reservation: "There's a bomb on the plane!" (1999, 2). "It seems bizarre on the face of it," Kushner writes,

> that something as rooted in culture as the utterance of inappropriate phrases or obscene words could be attached to organic disease. What in some societies are viewed as outrageous curses are seen as inoffensive in others. Moreover, even within the same societies words

> lose their offensive connotations over time. What is most interesting about coprolalia in Tourette's sufferers is that they invoke the most inappropriate curse of their particular times and cultures. (7)

Here, unlike Miller, Kushner situates the behaviors of Tourette syndrome by means of narrative within its cultural context.

Another historical approach is the case histories of Sacks. Here Sacks narrates the "discovery" of the condition in *The Man Who Mistook His Wife for a Hat.*

> In 1885, Gilles de la Tourette, a pupil of Charcot, described the astonishing syndrome which now bears his name. "Tourette's syndrome," as it was immediately dubbed, is characterised by an excess of nervous energy, and a great production and extravagance of strange motions and notions: tics, jerks, mannerisms, grimaces, noises, curses, involuntary imitations and compulsions of all sorts, with an odd elfin humour and a tendency to antic and outlandish kinds of play. (1987, 92)

Sacks offers the history of the condition itself, both in the lives of patients (case histories) and within historical cultures. Both Kushner and Sacks situate Tourette syndrome through narrative, in a manner similar to the way Peirce's index situates discourse.

A further definition can be found in fictional and literary treatments of Tourette syndrome, the ways it has been able to shape language, narrative, and the very subject of experience and discourse. Later in this chapter, I examine Jonathan Lethem's powerful novelistic portrayal of Tourette syndrome, but here I offer the description of a young girl with Tourette syndrome from the novel *Icy Sparks* by Gwyn Hyman Rubio. Rubio did not suffer from Tourette syndrome but from epilepsy as a child.

> *More'n anything in the world,* [grandpa's] last sentence, took over my mind. *World,* his last word, loomed there, large and greedy. The *world* was big, and he loved me *more'n anything* in it. If I didn't repeat *world,* it would grow larger and larger. Soon it would expand and extend itself from the top of my head to the tips of my toes. Like an enormous parasite, it would live in my body, change into a breathing, thriving, eating *world,* and devour me. So I had to say it. Saying *world* would diminish its power. *Pay homage to the word,* my thoughts told me, *and the world will be satiated.* (1998, 28–29)

Neither Rubio nor Lethem suffer from Tourette syndrome even though they created narrators—speaking subjects—who do. Here, language overwhelms the world, virtually *negates* it; the mind does not "grasp" things, as Saussure said the mind did (1983, 229); rather, words take over Icy's mind.

My last description is by Emma Morgan, a poet—an actual speaking subject—who herself suffers from Tourette syndrome. In her poem, the juxtaposition of a symptom of Tourette syndrome—kissing a wall—and a relative at a family reunion recognizes as equivalent different *kinds* of fact. "You see, the way it is with Tourette's," she writes,

> is that the symptoms change constantly and unpredictably, so that an unsuspecting Touretter might wake up one day to find that a tic she's had for years has totally vanished. Another day, she might acquire three new ones which could last a few days or weeks or even decades. Some tics seem to rotate according to a pattern, somewhat like the cycles of the moon. Others are more like the Halley's comet. One has to wait until they come around again. . . .
>
> I wrote "Complex Motor Tic" because I was so fed up with my newest addition shortly after its arrival.
>
> Complex Motor Tic
>
> I meet the wall
> with a brush of lips
> touch with side of face
> gauging precision of pressure
> like a well-dressed aunt
> at a family reunion,
> protecting the contour of makeup
> from a child's puckered face
> Sometimes I get it right
> in one try
>
> It's my latest tic
> How inconvenient,
> I think while driving
> holding out 'til traffic light turns red
> so I can nuzzle in safety
> with my steering wheel (1995, 5–6, 12)

In this poem—and elsewhere, as we will see—a variety of complex phonic tics are displayed: emotionally charged language; repetition of sounds (rhymes, alliteration); repetition of words; rhythms of language; sounds—even word-sounds—that are simply spooky to listen to. Above all, the complex phonic tics of Tourette syndrome—like the complex motor tic Morgan describes, the apparent kissing of objects in the world—are *uncanny:* here, the uncanniness of treating an object like a person. The uncanniness of Tourette syndrome is related, I think, to two aspects of the neurological description that Miller presents. The first is that in its complex forms, the tics of Tourette syndrome seem to present *meaningful gestures* and *meaningful sounds.* The second is that the simple and complex tics of Tourette syndrome *can be mimicked.*

The second of these facts seems peculiar once one thinks about it. Nicholas Cage in *Matchstick Man,* the voice of the Devil in *The Exorcist,* the psychiatrist's son in *What About Bob?* all *imitate,* in their cinematic representations, the uncanny voices of Tourette syndrome. Tourette syndrome is closely related to automatic calling phenomena in mammals, and especially in primates. Terrence Deacon has written a wonderful book called *The Symbolic Species: The Co-Evolution of Language and the Brain* in which (among other things) he discusses "the interactions between intentional motor systems and automatic calling tendencies" in primates, including humans. One example he offers is an observation by Jane Goodall, the woman who spent a lifetime observing chimpanzees in Africa. "Chimpanzees," Deacon notes,

> often produce food calls when they come upon a new food source. This stereotypic call attracts hungry neighbors to the location, often kin who are foraging nearby. Goodall recounts one occasion where she observed a chimp trying to suppress an excited food call by covering his mouth with his hand. The chimp had found a cache of bananas she had left to attract the animals to an observation area, and as she suggests, apparently did not want to have any competition for such a desirable food. Though muffling the call as best he could with his hand, he could not, apparently, directly inhibit the calling behavior itself. (1997, 244)

I mention this here to suggest that the powerfully *automatic* nature of such primate, primal cries can, at least among us human primates, be imitated and put to other uses. Our ordinary "primal cries," like those of our closely related primate, the chimpanzee, are also social calls: laughter and sobbing. Television laugh tracks really work: humans have a tendency to laugh in response to other laughter; and we have the tendency to sob in the presence of sobbing people. But it is very difficult to imitate laughter: trained actors can do it, but most "forced laughter," like forced smiles on our faces, rings false. Smiles are a good case in point: true smiling is a combination of intentional and automatic facial muscle patterns (see McNeil 1998), and fake smilers—Richard Nixon comes to mind, but also some of our more distant relatives (like Morgan's well-dressed aunt)—only present the voluntary muscle patterns. Trained actors learn to use involuntary muscles as well. Perhaps this is the strength of method acting (which is related to the discussion of memory in the next chapter). But the phenomena of Tourette syndrome are more easily imitated than smiles, even if, like the chimp's cry and some of our laughter, when they are real they cannot be fully suppressed.

The first of the facts of Tourette syndrome I listed, that it seems to present meaningful phenomena, is also particularly striking—and it is primary, even though I am discussing it second. It is striking on two levels. First of all, it powerfully confuses biological and cultural phenomena: as Kushner says, it confuses something rooted in culture with something rooted in physiology. As I have already suggested, Tourette syndrome is powerfully rooted in the physiology of brain chemistry. Equally interesting, I think, is that it combines motor/gestural phenomena with phonic/linguistic phenomena, two things which, intuitively, we think of as very different. But there are close connections between motor activity and linguistic activity: think for a moment of the remarkable fact that we can, automatically and unconsciously, make all the physical changes necessary to speak and be understood while we are eating crackers. Philip Lieberman, whose book *Human Language and Our Reptilian Brain* I cite later, has closely studied the relationship between motor skills and human language.

Physiology and Culture

Let me turn more directly to Tourette syndrome by finishing the quote from Oliver Sacks I already began, since the elegant description of Tourette syndrome he wrote several decades after the first edition of *Awakenings* remains an important and powerful description of the ways in which the physical and physiological bases of Tourette syndrome establish themselves within the hierarchy of levels of understanding I describe in *Intangible Materialism*. His description of "the affective, the instinctual and the imaginative life" catalogs these levels. "In 1885," he writes,

> Gilles de la Tourette, a pupil of Charcot, described the astonishing syndrome which now bears his name. "Tourette's syndrome," as it was immediately dubbed, is characterised by an excess of nervous energy, and a great production and extravagance of strange motions and notions: tics, jerks, mannerisms, grimaces, noises, curses, involuntary imitations and compulsions of all sorts, with an odd elfin humour and a tendency to antic and outlandish kinds of play. In its "highest" forms, Tourette's syndrome involves every aspect of the affective, the instinctual and the imaginative life; in its "lower," and perhaps commoner, forms, there may be little more than abnormal movements and impulsivity, though even here there is an element of strangeness. . . . It was clear to Tourette, and his peers, that this syndrome was a sort of possession by primitive impulses and urges: but also that it was a possession with an organic basis—a very definite (if undiscovered) neurological disorder. (1987, 92)

Tourette syndrome, Sacks noted a few years later in *An Anthropologist on Mars,* is "characterized, above all, by convulsive tics, by involuntary mimicry or repetition of others' words or actions (echolalia and echopraxia), and by the involuntary or compulsive utterances of curses and obscenities (coprolalia)" leading some to "strange, often witty" associations, others to "a constant testing of physical and social boundaries," and still others to "a constant, restless reacting to the environment, a lunging at and sniffing of everything or a sudden flinging of objects" (1995, 77–78). As this suggests, Tourette syndrome inhabits the juncture between biological formations and cultural formations, between the motor tics of Tourette syndrome—squinting, tapping, arm waving, sticking out the tongue, even licking objects—and its phonic tics—clearing the throat, sniffing, barking, repeating verbal sounds, rhymes, puns, shouting obscenities. There is a strange energy to Tourette syndrome that Sacks describes throughout his work and that, as I note later in this chapter, Lethem embodies in his novel *Motherless Brooklyn.*

What fascinates me about this syndrome—as it does Sacks, Lethem, and even David Morris in his powerful study of late twentieth-century illness, *Illness and Culture in the Postmodern Age*—is the continuity it presents between motor and verbal activity. Experimental work on the neurology of language, as outlined by Lieberman, argues forcefully and persuasively for tight connections between motor activity and language skills by focusing on the seat of vertebrate motor activity in the subcortical basal ganglia.[5] F. A. Middleton and P. L. Strick argue that "the cerebellum and basal ganglia should no longer be considered as purely motor structures" but instead involved "in cognitive processes" (1994, 460; see also Damasio for an extended argument about the role of "emotion, feeling, and biological regulation in human reason," 1994, xiii); and Lieberman even suggests that the evolution of hominid upright walking is closely connected to the evolution of language (2000, 143, 151; see also Mithen 2006, 159–68). This connection between body and language—between seemingly immaterial cognitive activity and our bodily life (a connection that both Michael Corballis and Merlin Donald make in evolutionary terms, examined in the next chapter)—is underlined in Tourette syndrome, which, by definition, essentially combines, in the words of a handbook on Tourette syndrome, "the presence of multiple motor tics (twitches) and one or more vocal tics (or noises)" (Robertson and Baron-Cohen 1998, 45).

At least since the Enlightenment, the connection between body and spirit has often been denied. In a defining instance, as we saw in chapter 2, René Descartes argued that language is the very sign of the immaterial soul in humankind and that automatic, mechanical phonic responses to experience, were they possible, would have nothing to do with meaningful language or meaningful gesture, so that, as he said, "it is not conceivable" to create a machine capable of "appropriately meaningful" verbal responses to experience (1985, 140). Like Descartes's machine, tics of Tourette syndrome, whether they are phonic or motor, respond to the world in a machinelike way without presenting any of the "appropriately meaningful" responses that Descartes describes. Yet the tics of Tourette syndrome convey meaning and provoke responses that raise questions about the ways in which the "appropriateness" of response is measured and the ways in which the materialities and meanings of discourse are bound together. "Tics," Sacks argues,

> can have an ambiguous status, partway between meaningless jerks or noises and meaningful acts. Though the tendency to tic is innate in Tourette's, the particular *form* of tics often has a personal or historical origin. Thus a name, a sound, a visual image, a gesture, perhaps seen years before and forgotten, may first be unconsciously echoed or imitated and then preserved in the stereotypic form of a tic. (1995, 81)

In a more scientific discourse, James Leckman and Donald Cohen describe "severe tic disorder as a model neuropsychiatric disorder that exists at the interface of mind and body" (1988, 9). Sacks's narrative descriptions in "Witty, Ticcy Ray" and "A Surgeon's Life," Leckman and Cohen's scientific accounts, and Lethem's first-person novelistic treatment offer a wide array of discussions of Tourette syndrome and allow for its being taken up, like the homologies of evolution and the sounds and elements of language themselves, to a host of differing ends. Tourette syndrome, then, situated "partway between meaningless jerks or noises and meaningful acts" at the "interface of mind and body," seems to take up the very materiality of language and underlines its materiality even as it preserves it *as* language. Thus, Dr. Carl Bennett, the subject of Sacks's most elaborate essay on Tourette syndrome, "A Surgeon's Life," notes that "it is just the sound [of particular words] that attracts me. Any odd sound, any odd name, may start repeating itself, get me going." "Echolalia," Sacks goes on, "freezes sounds, arrests time, preserves stimuli as 'foreign bodies' or echoes in the mind, maintaining an alien existence, like implants. It is only the sound of the words, their 'melody,' as Bennett says, that implants them in his mind; their origins and meanings and associations are irrelevant" (1995, 88–89).

Such frozen and arrested sound is a resource for poetry, its rhymes, alliteration, its "melody," even (or especially) the odd sense of the impersonalness of its most intimate references. Nietzsche's *The Birth of Tragedy* offers a fine meditation on the impersonalness of lyric poetry (1967a, 49; see epigraph for this chapter). In *The Birth of Tragedy,* Nietzsche repeatedly associates lyric poetry with "primordial" existence and even the "primal cry" that Greimas mentions. In Dionysian art, he writes, "we are pierced by the maddening sting" of the pains of

> primordial being itself just when we have become, as it were, one with the infinite primordial joy in existence, and when we anticipate, in Dionysian ecstasy, the indestructibility and eternity of this joy. In spite of fear and pity, we are the happy living beings, not as individuals, but as the *one* living being, with whose creative joy we are united. (1967a, 104–5)

Nietzsche came to disavow the exuberance of this writing, but I suggest that the impersonal energies he sees called upon and transformed in lyric and tragedy may well be associated with primordial, reptilian brain structures. Gilles Deleuze describes a version of this impersonalness in Nietzsche when he argues that, for Nietzsche, "a phenomenon is not an appearance or even an apparition but a sign, a symptom which finds its meaning in an existing force. The whole of philosophy is a symptomatology, and a semeiology" (1983, 3). It is the genius of art to apprehend impersonal phenomena—perhaps even unintentional phenomena—as meaningful. Similarly, Sacks argues in the 1982 epilogue to *Awakenings* that "Nietzsche, almost alone of philosophers, sees philosophy as grounded in our understanding (or misunderstanding) of the body, and so looks to the ideal of the Philosophic Physician" (1999, 279). Thus Nietzsche describes "the catharsis of Aristotle" as a "pathological discharge," which philologists are not sure "should be included [either] among medical or moral phenomena" (1967a, 132). While I do not argue that poetry is in any way an "impersonal" medical condition, the physiological condition of Tourette syndrome sheds light on its working and its power.[6]

A striking example that makes the seeming unintentional impersonalness of meaning and poetry its very theme is D. H. Lawrence's poem "Tortoise Shout." This poem articulates the sexual cry of a male tortoise, its "tortoise eternity, / Age-long, reptilian persistence" (1964, 365). Before offering a Lawrentian baroque sexual allegory, the poem reduces itself, so to speak, to sounds that are almost unintelligible: "This last / Strange, faint coition yell / Of the male tortoise at extremity, / Tiny from under the very edge of the farthest far-off horizon of life" (366).

> A far, was-it-audible scream,
> Or did it sound on the plasm direct?
>
> Worse than the cry of the new-born,
> A Scream,
> A yell,
> A shout,
> A paean,
> A death-agony,
> A birth-cry,
> A submission,
> All tiny, tiny, far away, reptile under the first dawn. (364)

"The Tortoise Shout" attempts to articulate—or, at least to describe—"deep," primordial meaning, a primal reptilian cry.

Poetry and Tourette Syndrome

Here, then, is precisely my thesis: that primitive resources of language most starkly apprehensible in the extremity and dysfunctionality—in the very *materiality*—of Tourette syndrome are a source of much of poetry's power. More specifically, human language, as Lieberman contends, "is overlaid on sensorimotor systems that originally evolved to do other things and continue to do them now" (2000, 1; see also 123, 156; and Deacon 1998, 298), and that poetry calls upon all the resources of the "functional language system," based on the subcortical or the reptilian brain as well as the neocortex, to achieve its power and its meanings. To make this argument, I reiterate Greimas's semiotic description of poetry, especially in the context of the evolution of language and cognition. Greimas notes that "at the moment of perception" a listener eliminates "about 40% of the redundancies of the distinctive phonological features unnecessary for the apprehension of meaning" (1972, 16; my translation). Lieberman describes such "highly redundant" processing of language in discussing the difficulties of following "even 'well-formed' speech recorded under ideal conditions." "We are generally unaware of these problems," he writes, "because we 'fill in' missing information, overriding acoustic phenomena that conflict with our internally generated hypotheses concerning what was *probably* said, and the probability involves a weighting of semantic and pragmatic information derived from parallel, highly redundant processing. Many of these phenomena can be explained if we take into account the distributed, parallel processing that appears to typify biological brains" (2000, 24–25). This description takes its place within Lieberman's larger argument of the redundant and overdetermined nature of Darwinian evolution. "Indeed," he writes, "speech perception is not a strictly 'bottom-up' process in which only primary acoustic or articulatory information is available to a listener. Many studies have demonstrated that what a listener 'hears' also depends on lexical and pragmatic information" (58). Such pragmatics, he argues, is not "logical" but "proximate": "evolution is a tinkerer, adapting existing structures that enhance reproductive success in the ever-changing conditions of life" (166; see also Lieberman 1998, 18–20; and Damasio 1994, 190). A chief example of this aspect of his argument is what he calls "motor equivalence," "the ability of animals and humans to accomplish the same goal using different muscles or different body parts" (2000, 39).[7]

In this context of the proximate logic of evolution, Greimas describes the manner in which poetry takes up the redundancies of language for other purposes. Just as ordinary listeners eliminate 40 percent of redundancies in ordinary uses of language, so "inversely," he argues, "the reception of the poetic message can be interpreted as the valorization of redundancies which become significative with the changing of the level of perception, valorization which would give rise to the apprehension of regularities . . . of sound, of connotation as it were, and not only of denotation" (1972, 16; my translation). By "valorization," Greimas means that superfluous redundancies of sound—but also redundancies of grammar or even semantics—come to constitute a level of meaning or a "meaning-effect" rather than simply perform as a (more or less redundant) vehicle for meaning that can be eliminated once meaning is communicated. Such meaning-effects are the *phenomena* of meaning: the felt sense of comprehension, the signifying whole beyond the individual elements of a sentence, for instance, or the logic of an argument, the genre of an extended discourse, the moral of a tale.[8] But the phenomena of meaning-effects include other felt senses discourse provokes, such as sadness, anxiety, fear, joy (see Schleifer 1987, xix). And poetry, Greimas argues, creates or provokes all of these effects by taking up and using—in Sacks's language, by freezing and arresting—the "disposable" material redundancies of language in ways that make them essential. In fact, this transformation of functional redundancies into aspects of discourse that create the impression of essential meaningfulness is another example of Greimas's description of the creation of disorder out of order I mentioned in chapter 2, where the "edifice" of language "appears like a construction without plan or clear aim" in which, for instance, "syntactic 'functions' transform grammatical cases by making them play roles for which they are not appropriate" (1983, 133).

This description of poetry, emphasizing as it does the phenomenal *materiality* of language, is at odds with the traditional opposition between matter and spirit. Again, Descartes makes this clear when he notes that "it may happen that we hear an utterance whose meaning we understand perfectly well, but afterwards we cannot say in what language it was spoken" (1985, 81). In *The World,* he uses this example to argue for a mechanical description of light:

> if words, which signify nothing except by human convention, suffice to make us think of things to which they bear no resemblance, then why should nature not also have established some sign which would

> make us have the sensation of light, even if the sign contained nothing in itself which is similar to this sensation? Is it not thus that nature has established laughter and tears, to make us read joy and sadness on the faces of men? (81)

Descartes posits a mechanical explanation for sensation, exactly what David Chalmers argues against. Even so, the Cartesian subject reads rather than participates in emotion: for Descartes, language is always a vehicle for meaning, never a provocation, a meaning-effect.

Moreover, his reference to the "joy and sadness on the faces of men" is particularly apt because it is probable that the neurobiology of emotions—which include joy, sadness, and a seemingly innate ability of primates to respond to faces—is closely connected to the strength and strange fascination of Tourette syndrome, its *situation* at the juncture of motor and verbal resources, between the intentional verbal meanings of discourse and its seeming unintended force.[9] That is, in its combinations of motor and phonic tics, Tourette syndrome uncovers redundancies that are often ignored. Lionel Essrog, the Tourettic narrator of Lethem's novel *Motherless Brooklyn,* says as much: "Tourette's teaches you what people will ignore and forget, teaches you to see the reality-knitting mechanism people employ to tuck away the intolerable, the incongruous, the disruptive—it teaches you this because you're the one lobbing the intolerable, incongruous, and disruptive their way" (1999, 43).

Early in the novel, Essrog introduces himself in ways that demonstrate how Tourette syndrome seems to depend on and emphasize (if not valorize) the materiality of language.

> Lionel, my name. Frank and the Minna Men pronounced it to rhyme with *vinyl.* Lionel Essrog. *Line-all.*
>
> Liable Guesscog.
>
> Final Escrow.
>
> Ironic Pissclam.
>
> And so on.
>
> My own name was the original verbal taffy, by now stretched to filament-thin threads that lay all over the floor of my echo-chamber skull. Slack, the flavor all chewed out of it. (7)

"Filament-thin threads." There is a strange hauntingness of the phonological and metaphorical language here: even the extended metaphor of "taffy" possesses a filament-thin materiality, and its combination of slackness and flavorlessness in the context of the almost metaphysical wittiness of the conceit presents almost intolerable incongruity.

These aspects of Tourette syndrome—its hovering between meaning and meaninglessness in the sounds and meanings of language, its revelations of the "reality-knitting" aspects of discourse, its playfulness and wit, its pathos and bathos—all these things, as Greimas says, are what poetry does as well, are sources and resources for poetry. "What is common to all [poetic] phenomena," he argues in a precise semiotic description,

> is the shortening of the distance between the signifier and the signified: one could say that poetic language, while remaining part of language, seeks to reachieve the "primal cry," and thus is situated midway between simple articulation and a linguistic double articulation. It results in a "meaning-effect," . . . which is that of "rediscovered truth" which is original and originary. . . . It is [an] illusory signification of a "deep meaning," hidden and inherent in the [phonological] plane of expression. (1970, 279; my translation)

The double articulation of language *is* the opposition between material signifier and immaterial signified, between the "distinctive phonological features" Greimas describes and the semantic wholes apprehended as meaning that do not seem, phenomenally, reducible to any part or even any precise combinations of parts. I spend some time focused on double articulation in the next chapter: the transformation of a sensuous sound into what Saussure calls the "sound-image" (1959) or the "sound pattern" (1983) of the signifier suggests the relationships among icon, index, and symbol. Indeed, this transformation situates the index between material sensation and seeming immaterial meaning, a kind of intangible material substance, like gesture pointing *toward* something.[10] Poetry, in Greimas's definition, attempts to create the illusion, the meaning-effect, that the signifier of the symbolic and communicative system of language becomes what cognitive neuroscience calls the "vocal signals" of primates (Emery and Amaral 2000, 174).[11] The vocal signals neuroscience describes, even when they make possible or manifest primate social organization, are themselves not structures of communication. They are primal cries in which the distance between signifier and signified, between sound and import, does not exist. Unlike the language Descartes describes in *The World,* the import of a vocal signal cannot "ignore" its material manifestation, the mechanical signals of primates.

The difference between a vocal signal and signifier—between mechanical noise and meaningful gesture (or between what Merlin Donald calls "mimicry" and "mimesis" [1991, 169])—might become clear in looking at a strange, even uncanny, phenomenon in English poetry, poems that rhyme words with themselves. This is described in the rhetorical term *rime riche*.[12] *Rime riche* seems to be a willful mimicking of the palilalia, repeating one's own words, that sometimes occurs in Tourette syndrome. In its mimicry, it often creates the illusion of uncontrollable, almost hysterical force. For instance, in a powerful short poem, "A Deep-Sworn Vow" (1919), W. B. Yeats rhymes "face" and "face."

> Others because you did not keep
> That deep-sworn vow have been friends of mine;
> Yet always when I look death in the face,
> When I clamber to the heights of sleep,
> Or when I grow excited with wine,
> Suddenly I meet your face. (1956, 152)

In a remarkable analysis, Sharon Cameron focuses on "the too-close bond between the two lines: 'Yet always when I look death in the face . . . Suddenly I meet your face" (1979, 218). The rhyming of the face of death and the face of the beloved enacts, I believe, the illusion of a primal cry Greimas is describing. In this operation of rhyming, a "sound" becomes a signifier not in order to mean something but to create the illusory (or phenomenal) identification of the beloved and the supernatural: looking at the face of death, Yeats encounters powerfully, seemingly automatically, the face of his beloved, and the wild disorder that occurs when, as Greimas says, "different dimensions"—in this case life and death—"can be at the same time recognized as equivalent" (1983, 82). Yet even while the poem creates the *illusion* of automatic, mechanical identification, it creates this illusion within the contexts of systems of rhyme and concepts of rhyme, which have no place in relation to the phenomenon of self-stimulating vocal signals.

In important ways, the phonic tics of Tourette syndrome seem to be *simply* mechanical, self-stimulating vocal signals. The fact that in clinical trials three decades ago both motor and phonic tics were suppressed by dopamine-blocking drugs—haloperidol, in early instances (see Kushner 1999, 133–43; he cites Stevens and Blachly 1966)—suggests its mechanical nature. Indeed, as Leckman, Riddle, and Cohen note,

> The basal ganglia and the substantia nigra are widely considered to be the neuroanatomical regions associated with a variety of movement disorders including Parkinson's disease, encephalitis lethargica, Huntington's disease, and tardive dyskinesia. Although the neuropathological correlates of TS remain to be fully established, the presence of abnormal movements in TS, suggestive neuropathological data, and a substantial body of pharmacological and metabolic data implicating neurochemical systems localized in these regions have led to the hypothesis that the pathophysiology of TS and related disorders may involve some dysfunction of these areas. (1988, 105)

Here, as in the neurobiology of emotion—which is, like Tourette syndrome, associated with subcortical regions of the brain—Tourette syndrome seems to realize itself in relation to our reptilian brain. Tourette syndrome, Sacks writes,

> like Parkinsonism and chorea, reflects what Pavlov called "the blind force of the subcortex," a disturbance of those primitive parts of the brain. . . . In Tourette's, where there is excitement of the emotions and the passions, a disorder of the primal, instinctual bases of behaviour, the disturbance seems to lie in the very highest parts of the "old brain": the thalamus, hypothalamus, limbic system and amygdala, where the basic affective and instinctual determinants of personality are lodged. (1987, 95–96)

Echoing, repetition, puns, punctuated language—erasing in its barks and noises the distance between the signifier and signified even as it excites the emotions and passions: this description of Tourette syndrome might help us to see some of the resources of language poetry attempts to "reachieve."

The most well-known aspect of Tourette syndrome, its coprolalic barking of obscenities, is tied up with the materiality of language—both its material soundings and its material neuroanatomic pathways. In fact, David Morris argues that in important ways, "midbrain and limbic structures [function] in the control of obscene words" (1998, 174). In this argument, Morris assumes that cognitive/expressive language *simply* originates in the neocortex and, for that reason, is distinctly "human." "Human speech," he writes, "however it developed during the long history of evolution, did *not* develop from the cries and vocalizations of nonhuman primates. Human speech differs fundamentally from animal cries in the sense that it proceeds from an entirely different region of the brain" (174).[13] Certainly neurological studies have shown, as both Lieberman and Deacon note, that "neocortical areas do not appear to regulate voluntary vocalizations in nonhuman primates; neither cortical lesions nor stimulations affects their vocalizations" (Lieberman 2000, 99; he cites MacLean and Newman 1988 and Sutton and Jurgens 1988; see also Deacon 1998, 235–36). And more strikingly, Jane Goodall observes that "chimpanzee vocalizations are closely bound to emotion. The production of a sound in the *absence* of the appropriate emotional state seems to be an almost impossible task for a chimpanzee" (1986, 125).

But, as might already be clear, I suggest that Morris is not altogether correct in his contention that all speech differs simply and fundamentally from the primal cries of primates.[14] In fact, even Morris suggests some aspects of primal discourse inhabit language when he cites studies that demonstrate that "aphasias that cripple or destroy normal speech, leaving patients unable to talk or write, sometimes preserve untouched the ability to swear like a sailor" (1998, 174; see also Pinker 1994, 301), and he argues that "an obscenity, after all, is more like a cry than a word. Or rather, it belongs to a special class of words that serve as the direct expression of emotion" (1998, 174). A friend and brilliant scholar who suffers from Tourette syndrome has warned me not to romanticize Tourette syndrome in the ways that R. D. Laing romanticized schizophrenia a generation ago. Morris, in this passage, comes close to such romanticization insofar as he suggests that the tics of Tourette syndrome are interpersonally expressive. There is, as Morris says, a class of words that serve in powerful ways to express emotion, but their use—or really "mention"—in Tourette syndrome is not expressive even if they can create the "effect" of expression. Still, the phonic "mention" in Tourette syndrome of sounds which function in verbal "use" most forcefully juxtaposes the biological and cultural formations inhabiting language that I mentioned earlier.[15] In his study, Morris argues that "the obscene achieves its apparently ineradicable place by weaving together powerful elements of our biology, psychology, and social life" (166). These "weavings" involve "old brain" subcortex as well as the neocortex, including the basal ganglia that Lieberman argues constitutes an essential part of the functional language system, and together—as elements of biology, psychology, and social life—they serve poetry.

Finally, the tics of Tourette syndrome, both motor and phonic, are closely associated with obsessive-compulsive behavior (see Gray 1994,

50; he cites Pauls et al. 1986; Kushner 1999, 205, 216). Such behavior blurs or suspends the opposition between intentional and involuntary actions insofar as it shapes itself in relation to context (see Kushner 1999, 197–99).[16] A host of scientific studies have demonstrated that 10 to 40 percent of patients subject to tic disorders "report *obsessional thoughts* and exhibit *compulsive behaviors*" (Leckman and Cohen 1988, 10; see also Towbin 1988), and Sacks's case histories also make this abundantly clear, as does Lethem's powerful fictional portrayal of Tourette syndrome. As I already suggested, the verbal manifestations of Tourette syndrome are often context sensitive (see also Kushner 1999, 2–3), taking the form of repetition of sounds, including echolalia, but also palilalia, "the repetition of the patient's own last word, phrase, or syllable" (Robertson and Baron-Cohen 1998, 21), and, as both Sacks (1987) and Lethem portray, they often take the forms of verbal rhymes and puns. Neurobiologists suggest that obsessive-compulsive behavior may be closely connected to grooming behavior in other mammals (Gray 1982, 443), and Lieberman argues that such behavior—regulated in the subcortical basal ganglia—is parallel to the syntax of language (2000, 87).[17] Thus, the very neurobiology of language—which includes "phylogenetically 'primitive' neuroanatomical structures found in the brains of 'lower' animals, such as the cerebellum and basal ganglia" (20)—suggests the opposition between the intentional activities of mind and the automatic activities of body is more complicated than we thought. Deacon notes that "even for humans, the essentially automatic and unconscious nature of many stereotypic calls [primal cries] causes them to erupt without warning, often before there is time to interfere with their expression. . . . This curious conflict between simultaneously produced intentional and unintentional behaviors offers a unique insight into the nature of language. The superimposition of intentional cortical motor behaviors over autonomous subcortical vocal behaviors is, in a way, an externalized model of a neural relationship that is internalized in the production of human speech" (1998, 244). This conflict is present, in notably different ways, in Tourette syndrome and in poetry.

The Power of Poetry

Even if Tourette syndrome arises from biophysiological grounds, it involves, as Morris suggests, elements of psychology and social life as well. Sacks notes that in the momentary freedom from Tourette syndrome when Dr. Bennett performs surgery, "one is seeing something at a much higher level than the merely rhythmic, quasi-automatic resonance of the motor patterns; one is seeing (however it is to be defined in psychic or neural terms) a fundamental act of incarnation or personation, whereby the skills, the feelings, the entire neural engrams of another self, are taking over in the brain, redefining the person, his whole nervous system, as long as the performance lasts" (1995, 98). Sacks is talking about the ways in which Tourette syndrome seems to be "suspended" momentarily by rhythmic activity in general and, in this specific case, by the art of surgery. Elsewhere, he presents Tourette syndrome in the opposite fashion. In *Musicophilia,* he cites the professional jazz drummer, David Aldridge, describing how "rhythm and Tourette syndrome have been intertwined from the first day I found that drumming on a table could mask my jerky hand" (2007, 228), and in "Witty, Ticcy Ray," he quotes Ray claiming that he cannot imagine his life without his tic: "I consist of tics—there is nothing else" (1987, 98). "He said," Sacks concludes, "he could not imagine life without Tourette's, nor was he sure he would care for it" (98). In both cases, Sacks depicts what the neurologists Cohen, Bruun, and Leckman describe in the preface to *Tourette's Syndrome and Tic Disorders: Clinical Understanding and Treatment.* Tourette syndrome, they say, "affects individuals at the core of their experience of themselves as being in control of their own movements, statements, and thoughts" (1988, xiii).

At the heart of this "core" is language. Tourette syndrome affects this core as it manifests a kind of "material language" in relation to what Sacks describes as "selfness" and its connection to the seeming intentionality of language. It is "often difficult for Touretters," Sacks writes, "to see their Tourette's as something external to themselves, because many of its tics and urges may be felt as intentional, as an integral part of the self, the personality, the will" (1995, 102; see also 1999, 109–11; 2007, 231–32). Early in *Motherless Brooklyn,* Lethem describes the progression from early motor tics in patients suffering from Tourette syndrome to phonic tics. (Clinicians note that motor tics usually start occurring at about age seven, while phonic tics usually occur between four and seven years later [Bruun 1988, esp. 23–25].) For a time, Lionel Essrog's ticcing took the form of kissing the other boys in the Brooklyn orphanage:

> The kissing cycle was mercifully brief. I found other outlets, other obsessions. . . . [Instead, I was] prone to floor-tapping, whistling,

> tongue-clicking, winking, rapid head turns, and wall-stroking, anything but the direct utterances for which my particular Tourette's brain most yearned. Language bubbled inside me now, the frozen sea melting, but it felt too dangerous to let out. Speech was intention, and I couldn't let anyone else or myself know how intentional my craziness felt. Pratfalls, antics—those were accidental lunacy, and more or less forgivable. Practically speaking, it was one thing to stroke Leshawn Montrose's arm or even kiss him, another entirely to walk up and call him Shefawn Mongoose, or Lefthand Moonprose, or Fuckyou Roseprawn. (Lethem 1999, 47)

Essrog's experience of unwilled intention, so to speak, is the problem of Tourette syndrome and, in a way, the problem of poetry. Words call themselves forth by sound and rhythm, by their performability, that—however "crazy" they seem—*feel* intentional.

In "Tradition and the Individual Talent," T. S. Eliot describes "the bad poet" as one who "is usually unconscious where he ought to be conscious, and conscious where he ought to be unconscious. Both errors tend to make him 'personal'" (1975, 43).[18] Another way to say this is to describe poetry as the articulation of the kind of language I describe throughout this chapter: the *material* articulations of language that gather up the power and emotions of seeming subcortical primal cries within its discourses in the fact that the very materials of language—its sensate-producing sounds, its physiological organization in relation to brain chemistry—also organize themselves in relation to the level of semiotics. "T. S. Eliot." Neuroscientists abbreviate Tourette syndrome as "TS." Could there be a link here? If there is, such a link is "Tourettic," the creation of disorder out of order: it traffics in the materiality of language, making material *sound* seem intentional and semiotically meaningful, making impulse seem conscious. That is, the mechanical echoings, repetitions, rhythms, and emotionally charged phonic tics of Tourette syndrome present themselves as meaningful sounds and, as perhaps obscenities themselves do, trigger or are simply associated with more or less automatic emotional responses. Behind this "link" between the sounds and energies of poetry and discourse is the assumption that lyric poetry does not only—or does not simply—"express" the poet but also articulates what I described earlier in relation to Nietzsche as "impersonal energies," articulations that are *apprehended* in particular conscious and unconscious manners.[19]

This is, I believe, at least a part of what Eliot—and in their different ways Nietzsche and Lawrence—mean by "impersonal" poetry and what Lethem repeatedly figures in *Motherless Brooklyn* as the "boiling" of language. "Beneath that frozen shell of sea," Lionel says,

> a sea of language was reaching full boil. It became harder and harder not to notice that when a television pitchman said *to last the rest of a lifetime* my brain went *to rest the lust of a loaftomb,* that when I heard "Alfred Hitchcock," I silently replied "Altered Houseclock" or "Ilford Hotchkiss." (1999, 46)

Language boils in poetry. Listen, for instance, to T. S. Eliot's "Ash Wednesday":

> If the lost word is lost, if the spent word is spent
> If the unheard, unspoken
> Word is unspoken, unheard;
> Still is the unspoken word, the Word unheard,
> The Word without a word, the Word within
> The world and for the world;
> And the light shone in darkness and
> Against the Word the unstilled world still whirled
> About the centre of the silent Word.
> O my people, what have I done unto thee. (1964, 90)

All the material resources of poetry are here: rhymes, alliteration, repetition, the puns of *rime riche,* unmarked quotation, incantation that almost barks. Signifier and signified approach one another to the point of signal, to the point of primal cry: "the Word without a word, the Word within / The world and for the world." In this poetry, an insistent material discourse pushes, almost Tourettically, toward a word without a word within the world whirling and still.

In this passage from "Ash Wednesday," Eliot gathers together the energies of language to make the unheard word whirl the world provokingly, startlingly, affectingly. "Tics," Sacks says, "are like hieroglyphic, petrified residues of the past and may, indeed, with the passage of time become so hieroglyphic, so abbreviated, as to become unintelligible (as 'God be with you' was condensed, collapsed, after centuries, to the phonetically similar but meaningless 'goodbye')" (1995, 81). Hugh Kenner describes the mechanism by which the signifiers of what he calls "post-Symbolist" poetry float, tic-like, above meaning. Citing a couplet from Shakespeare's *Cymbeline,* "Golden lads and girls all must / As chimney-sweepers, come to dust," he describes the "magic" that "irradiates the stanza" so that "we, the heirs of Mallarmé and Valéry and Eliot, do not simply pass over 'golden' but find it richly Shakespearean." Moreover, he describes what Sacks would call the hieroglyphic abbreviations of this poetry: he notes that in Shakespeare's Warwickshire "golden boys" is the name for dandelions, and they are called chimney-sweepers when they go to seed. "We may want to say," he argues, "that Shakespeare wrote about happenings in the world, the world that contains mortal men and sunlight and dandelions, and that a post-Symbolist reading converts his work into something that happens in the language, where 'golden' will interact with 'dust' and 'wages' and 'lads' and 'girls' and 'chimney-sweepers,' and where 'dust' rhymes with 'must,' mortality with necessity" (1971, 122, 123).

Greimas also notices this phenomenon in modern poetry. Modern poetry aims, he argues, at "'abolishing syntax,' that is to say, diminishing as much as possible the number of functional messages" in order to iterate "a certain number of semic categories" that constitute poetic communication (1983, 153–54). For Greimas, the semic categories are the distinctive features of meaning or semantics homologous in organization to the distinctive phonological features about which he wrote. Semic categories, he argues, "probably contain the universals of language" (Greimas and Courtés 1982, 278), by which he means categories organizing experience shared by all people.[20] Many of these categories—of spatial relationships, of emotional response to experience, of motor balance in the world—neurology suggests, depend in important degree on subcortical regions of the brain. And many of them are mimicked, disrupted, incongruously enacted in the motor and phonic symptoms of Tourette syndrome.[21] In Greimas's analysis, then, modern poetry aims at creating a semantics that is seemingly without syntax, which is to say a semantics in which the opposition between word and thing—between the two articulations of language or between linguistic and motor activity—pushes toward the "rediscovered truth" of a simple rather than a double articulation.

The verbal tics of Tourette syndrome are, like the modern poetry both Kenner and Greimas describe (in their remarkably different idioms), something that happens in the language that disrupts and dislodges by means of the intolerable, incongruous, and disruptive materiality of language. Such materiality is no simple metaphor: it is literally materially inscribed within our brain structure and brain chemistry, powers we share with primates and other mammals that bubble forth to disturb and affect us in the impulsive utterances in Tourette syndrome and that are gathered up to one degree or other in poetic discourses. Poetry enacts the "intentional materiality" of discourse we can hear in Tourette syndrome and in T. S. Eliot. Listen to the end of section 5 of "Ash Wednesday."

> Will the veiled sister between the slender
> Yew trees pray for those who offend her
> And are terrified and cannot surrender
> And affirm before the world and deny between the rocks
> In the last desert between the last blue rocks
> The desert in the garden the garden in the desert
> Of drouth, spitting from the mouth the withered apple-seed.
> O my people. (1964, 91)

Eliot's feminine rhymes, "slender," "offend her," "surrender," slide "her," the unnamed Mary, into an unaccented schwa, an unheard word, a vocal signal. Like Lionel's "Altered Houseclock," Eliot's echoed sounds, repeated words, even the spondee of "the last blue rocks" slowing down the line, or his archaic "drouth," mispronouncing our English to rhyme with *mouth,* all play with sounds and meanings to create an illusion of articulated but not quite apprehensible meaning hidden and inherent in the seeming signals of poetry.

Such play is, of course, more than play. As Dr. Bennett told Oliver Sacks, Tourette syndrome is

> not gentle. . . . You can see it as whimsical, funny—be tempted to romanticize it—but Tourette's comes from deep down in the nervous system and the unconscious. It taps into the oldest, strongest feelings we have. Tourette's is like an epilepsy of the subcortex; when it takes over, there's just a thin line of control, a thin line of cortex, between you and it, between you and that raging storm, the blind force of the subcortex. One can see the charming things, the funny things, the creative side of Tourette's, but there's also that dark side. (1995, 100)

Poetry, too, sometimes taps the oldest, strongest feelings we have, even as it puts up a thin line of control. Moreover, it does so in uses of language that, again at times, picks up the *materials* of language that Tourette syndrome articulates, cursing, rhyming, patterning discourse—above all *embodying* discourse—to provoke and arouse elements of our oldest emotional lives.

4. The History of the Hand: Peirce's Index, Attention, and the Power of Narrative

> The experience [of grooming] is both physical sensation and social intercourse. A light touch, a gentle caress, can convey all the meanings in the world: one moment it can be a word of consolation, an apology, a request to be groomed, an invitation to play; on another, an assertion of privilege, a demand that you move elsewhere; on yet another a calming influence, a declaration that intentions are friendly. Knowing which meaning to infer is the very basis of social being, depending as it does on a close reading of another's mind. In that brief moment of mutual understanding in a fast-moving, frenzied world, all social life is distilled in a single gesture.
>
> —*Robin Dunbar,* Grooming, Gossip, and the Evolution of Language

Chapter 3 examines the relationship between physiological materialism of voice and the formal affective elements of poetry; this chapter examines the relationship between the evolutionary materialism of the hand and the elements of literary narrative. Near the end of the preceding chapter, I suggested—almost hidden in a footnote—that Darwinian accounts of evolution are "similar" to linguistic notions of the arbitrary nature of the sign and to Oliver Sacks's description of the conventionality of space-experience as more or less "convenient." The similarity, as I noted there, is between the accidental and more or less arbitrary nature of the "proximate" causes of evolution (see Mayr 1982) and the comprehension of the elements of language and meaning as more or less arbitrary. I even ended the note with the assertion that Philip Lieberman sounded a lot like Claude Lévi-Strauss in his argument that "evolution is a tinkerer, adapting existing structures that enhance reproductive success in the ever-changing conditions of life" (2000, 166). In *The Savage Mind,* Lévi-Strauss likens mythical thought to "a kind of intellectual 'bricolage'" (1966, 17), which takes up whatever is at hand for its work. Still, despite this seeming similarity between the arbitrary organization of language and the explanations of evolution in terms of proximate causes, biological adaptations are not quite as *purely* relational or as *purely* nonreferential as language is sometimes described in Saussure or Greimas (even if a student of evolutionary neurology like Terrence Deacon offers a finely textured account of referentiality in relation to brain function and learning that comes close to accounts of referentiality in many continental linguists [1998, 59–92]).

Rather, even though evolution responds to historical accidents of one sort or another—changing climates, opening survival niches (see Weiner 1995), catastrophes of meteors or even "shifts [of] natural selection into the quantum realm" (McFadden 2000, esp. 270)—the "convenient" and seemingly "arbitrary" nature of a primate's normal sense of existence in time and space is always *contingently arbitrary,* which is to say it is dependent upon more or less accidental (and thereby arbitrary) environmental or "proximate" factors, but the actual evolutionary responses to those factors are motivated rather than arbitrary. As Schrödinger says, evolution creates order out of order. This is so even if the primate sense of time and space can be contrasted with the very different spatial and temporal experiences of dogs or frogs or even ticks (see Hayles 1991, 76–78): it is contingent, that is, in relation to the "ever-changing conditions of life." Thus, when Sacks describes the "violently deforming forces" to which his postencephalitic patients are subject that "drive" them to "misperformance and miscognition" in experiences in which "normal" spatial and temporal relationship do not function (1999, 344–45; see also Sacks's discussion of spatial experience for people with Tourette syndrome in *An Anthropologist from Mars* [1995, 83]), he is suggesting human (and even primate) "universals" of spatial and temporal apprehension—the "universals" of meaning, experience, and language—that are governed by contingent adaptive needs particular to primates. "Eating fruit," Emery and Amaral argue,

> is a relatively simple task for most primates. Good color vision is required to locate specific types of fruit and to assess the level of ripeness or toxins that may be present. Olfaction and taste are also important indicators of the palatability of food. Highly distributed resources such as ripe fruit require a highly developed spatial memory system to remember where a previously encountered desirable or plentiful food source is located within a forest environment. A fine level of dexterity may be required to reach fruits in the high branches of trees, and fine manipulative ability may be required for removing the skin and seeds of some fruits. (2000, 170)

Fruit eaters, they also argue, "are usually highly social," since "the majority of fruit-eating primates eat green or bitter fruits, which are plentiful and clumped in large resources, thereby enabling many animals to feed in one tree," which thus "provides increased opportunities for social interaction" (170). Here, then, they argue for the adaptiveness of dexterity, fine vision, spatial sense, and communicative systems based upon the arbitrary contingencies of adaption to an ever-changing world.

They also argue that all of these adaptations use and modify subcortical resources of the brain, especially the amygdala. The "universal" primate and human sense of space—our performance and cognition of it, particularly including the misperformances in some of the symptoms of Tourette syndrome—offers a way of bringing together the "cultural formations" of the negative structures of semantics as I have described them and the biological formations of Darwinian materialism by emphasizing the *constraints* governing evolution. Most important, I think, to the sense of primate space are our senses of touch: they make the space of our experience, first and foremost, a social space—a space for faces and voices as well as touch. Face, voice, and touch—and the shared sense of pain I explore in chapter 5—offer the possibility of emphasizing an understanding of materialism that can, in fact, be accommodated by semiotic accounts of meaning and experience. For this reason, I examine in this chapter the place of materiality of evolutionary social gesture in attention and meaning, just as I examined in the preceding chapter the place of meaningless material sounds in poetry and meaning. That is, as chapter 3 examined the materiality of voice by focusing on the physical chemistry and neurobiology of Tourette syndrome and their homological relationships with poetry, this chapter focuses on the more or less intentional activities of the human hand and their homological relationships with narrative. To this end, I examine the relationships among gesture, attention, and the narrative power of poetry in the context of Peirce's indexical semiotics. First, though, I want to touch on an early articulation of systems theory. Chapter 2 pointed toward systems theory in its examination of biological systems, and here I examine this theory more closely in relation to the functioning and "systematic" evolution of the human hand. In *What Is Life?* Erwin Schrödinger offers an early version of systems theory when he distinguishes between the laws of physics and the laws governing life processes. He asserts that the "statistical mechanism" of the laws of physics produces "order from disorder," while those governing life processes produce "order from order" (1992, 80), and in chapter 2, I also suggested the ways that Peirce's semiotics pointed toward "the nonmechanistic conclusions" implicit in "the probability and statistical revolution that forms the base on which twentieth-century science has been erected" (Depew 1994, 124).

In his distinction, Schrödinger describes two global hierarchical levels of organization, "nested" organization that has come to be described as systems theory. Recent work in information theory, semiotics, and—that world of fantastic speculation—quantum physics has suggested a third global hierarchical level, that of information, semiotics, and cognition, which I focus on in relation to pain in chapter 5. Thus, in 1948, not long after Schrödinger's work, Norbert Wiener proclaimed in *Cybernetics* that "information is information, not matter or energy. No materialism which does not admit this can survive at the present day" (1961, 132). I have described this third level as the production of "disorder from order" insofar as information systems are "irreducibly nonsimple" (to use a phrase from Jacques Derrida 1982, 13). Thus, as I noted earlier, Greimas perhaps best articulates this state of affairs in *Structural Semantics* when he argues that the "edifice" of language "appears like a construction without plan or clear aim" in which, for instance, "syntactic 'functions' transform grammatical cases by making them play roles for which they are not appropriate; entire propositions are reduced and described as if they behaved like simple adverbs" (1983, 133). A good example of what he describes is that it is always possible that a sentence can be "transformed" into a noun: "'I like Ike' is my favorite sentence." Such nonsimplicity, I suggest, characterizes attention insofar as attention itself *plays* hierarchically and self-reflectively between world and self, gestures and meaning, similes and contrary-to-fact assertions, and plays among Peirce's icons, indexes, and symbols. "I like Ike," for instance, is a remarkable example of the play among the iconic sensuousness of language conceived as sound, its indexical referentiality, and its symbolic meaning. It is Roman Jakobson's example of "a paronomastic image of a feeling which totally envelops its object" (1987, 70), which is an apt description of the kinds of affective language I described in chapter 3.[1] Recent studies of biology by Fritjof Capra and Johnjoe McFadden (among others) have suggested, as Capra says, that "in the emerging theory of living systems the process of life . . . is identified with cognition, the process of knowing" (1997, 168). "The simplest organisms," he writes, "are capable of perception and thus of cognition" (170). In this way, cognition is based—it is *nested*—upon biological evolution.

Cognition, moreover, can be shared, and Merlin Donald has argued that the evolution of shared cognition—the evolution of language and consciousness—was achieved by means of "the standardization of mimetic performance" in gestures (1991, 220). In a more recent study, Donald describes the way that "gesture requires splitting attention, remembering relationships, self-supervision, rehearsal, and metacognitive review" (2001, 145). Such metacognitive review—"knowing that you know"—is what Antonio Damasio describes as "the simple and standard dictionary definition of consciousness—an organism's awareness of its own self and surroundings" (1999, 4). Damasio goes on to describe this knowledge as "the presence of you" that "is the feeling of what happens when your being is modified by the acts of apprehending something" (10).[2] "Knowing that you know" is, of course, a remarkable and cryptic articulation of the self-reflective hierarchy of systems. Damasio argues—in great neurological detail—the systematicity of consciousness by analyzing the fact that "consciousness and emotion are *not* separable" (17). In his neurological analysis, he distinguishes between "core consciousness" (that is "not exclusively human; and it is not dependent on conventional memory, working memory, reasoning, or language" [16]) and "extended consciousness" (that is a "complex biological phenomenon" with "several levels of organization" evolving "across the lifetime of the organism" [16]).[3] Core and extended consciousness are themselves systematically organized.

The Evolution of the Hand

I suggest that attention and consciousness are evolutionarily, experientially, and above all systematically tied up with gesture, and specifically with gestures of our hands (as opposed to the primal cries of Tourette syndrome, which have little or nothing to do with attention and consciousness). To gesture is to mean.[4] And more: to gesture is to make self-reflective comprehensions of meaning possible. Frank Wilson passionately argues that "any theory of human intelligence which ignores the interdependence of hand and brain function, the historic origins of that relationship, or the impact of that history on developmental dynamics in modern humans, is grossly misleading and sterile" (1999, 7). Wilson asks what is the hand itself and concludes that there are "three quite different perspectives on the role of the hand in human life": a "biomechanical and physiological perspective"; an "anthropological and evolutionary perspective" that includes its "central role in human life and survival"; and a "neurobehavioral and developmental perspective" that describes "the dynamic interactions of hand and brain . . . and how that process relates to the unique character of human thought, growth, and creativity" (10). These three perspectives correspond to the levels of analysis examined in chapter 2: mechanical physics, interactive biology, and dynamic semiotics. Later in his study, Wilson makes this clear in terms of "a fundamental premise of Darwinian thought, that structure and function are interdependent and co-evolutionary. The brain keeps giving the hand new things to do and new ways of doing what it already knows how to do. In turn, the hand affords the brain new ways of approaching old tasks and the possibility of undertaking and mastering new tasks. That means the brain, for its part, can acquire new ways of representing and defining the world" (146).

On the biomechanical level, Wilson nicely traces the development of the primate hand from paws and the human hand from primate hands.[5] Among primates, "the need to cling to trunks and small branches favored a tactilely refined hand, and one whose movement and *control* repertoire had impressive range: from delicate, dissecting movements of individual digits to powerful grasping movements used both in sustained holding (prolonged suspension) and in the sequences of quick hold and quick release movements used in swinging and jumping" (28). The human hand developed with evolutionary changes in the wrist and hand—change in the ulnar bone in the forearm opposite the thumb—which allowed opposition of the fourth and fifth fingers with the thumb. This opposition, not possible for apes, permitted

> a stick to be seated tightly in the hand and oriented along the axis of the arm, so that the swinging radius of arm-plus-stick (and, therefore,

> the force of a blow) increases. . . . [Once this] "squeeze" grip was introduced into the hominid hand, no adversary or prey in the same weight class was safe in its presence without being unusually fleet-footed, hard-headed, or thick-skinned. (27)

A stick, like Peirce's "index"—or even the index finger itself, which in Latin was called *demonstratorius* (Napier 1993, 23)—extends the axis of the arm as well, the orientation from the subject of experience, that embodies, materially, Wilson's third level, the interactions of hand and brain. But "a second effect of ulnar opposition" Wilson describes is the "improved precision grip, in which small objects are manipulated between the fingers without contacting the palm. The hand with this biomechanical ability would have been able not only to wield a large club and to manipulate and throw stones but also to use all five digits in the fine control of small objects" (27–28).

In these ways, Wilson describes the physiology and the adaptiveness of the hand. The interaction of hand and brain is, first of all, a function of the social nature of the species. Thus, Robin Dunbar has argued that the major correlation to the increased size of the brain in primates—and especially the significant increase in the size of the neocortex ("80 per cent of the total brain volume in humans" as compared to "30 to 40 per cent in most mammals" [1996, 62])—is social interaction. He plotted the size of the neocortex in relation to the stable group size of seventy primate species and two nonprimate mammals. He found the larger the group, the larger the neocortex; in the nonprimate mammals, he found a large neocortex in the highly social vampire bats and a very small neocortex in solitary carnivores (65–66; see also Wilson 1999, 38–39). "Neocortex size," Wilson concludes from Dunbar's argument, "is a reliable predictor of group size because intelligence is mainly *social* intelligence; the more people there are to keep track of, the greater the complexity of relationships to be kept in mind and orchestrated, and the more time which must be spent maintaining coalitions" (39). The hand is important to all this, Dunbar argues, because the representative social activity among primates is grooming. "There is more to grooming than just hygiene," he says, "at least in the monkeys and apes. For them, it is an expression of friendship and loyalty" (1996, 21). "A grooming partner," he continues, "is something special, someone who deserves particular attention, who should be supported in moments of need, on whose behalf the taking of risks is warranted" (22). Dunbar also argues that grooming has the "ability to induce a state of relaxation and mild euphoria. This, if you like, is the reinforcer that makes monkeys willing to spend so much time in what would otherwise seem a pointless activity. Even though grooming ensures that the fur is cleaned and the skin kept free of debris and scabs, the time devoted to it by species like baboon, macaques and chimpanzees far exceeds that actually needed for these simple purposes" (38).

The kind of hands-on caretaking is nicely represented in a powerful poem by Hart Crane, "Episode of Hands."

> The unexpected interest made him flush,
> Suddenly he seemed to forget the pain,—
> Consented,—and held out
> One finger from the others.
>
> The gash was bleeding, and a shaft of sun
> That glittered in and out among the wheels,
> Fell lightly, warmly, down into the wound.
>
> And as the fingers of the factory owner's son,
> That knew a grip for books and tennis
> As well as one for iron and leather,—
> As his taut, spare fingers wound the gauze
> Around the thick bed of the wound,
> His own hands seemed to him
> Like wings of butterflies
> Flickering in sunlight over summer fields.
>
> The knots and notches,—many in the wide
> Deep hand that lay in his,—seemed beautiful.
> They were like the marks of wild ponies' play,—
> Bunches of new green breaking in hard turf.
>
> And factory sounds and factory thoughts
> Were banished from him by that larger, quieter hand
> That lay in his with the sun upon it.
> And as the bandage knot was tightened
> The two men smiled into each other's eyes. (1966, 141)

In this encounter of hands, self and other are organized around touch and gesture, and core and extended consciousness play together like springtime ponies. Part of that feeling, here, is that hands *know.* They know grip of book and iron—in the cyborg knowing Katherine Hayles describes, where a blind man's cane extends the body out into the world of things (1999, 84–85)[6]—and they also respond to the consciousness of other minds, other feelings. In this poem, hands and feelings and attention itself are confused and shared in the oddness of small hands holding the larger hand. More specifically—more linguistically—the antecedents of the poem's masculine pronouns confuse themselves so that the subject of the first three stanzas, the worker with his wounded hand—or at least the first two and a half, since it is hard to tell whose hands "seemed . . . like wings of butterflies"—somehow is transformed into the small-handed owner's son, who is the subject of the last stanzas, holding what seems the hand of a friend.

Indexicality and Narrative

This confusion of intertwining hands and pronouns—this work of the hand, helping us to apprehend and experience the world—also allows us to comprehend the awareness of others and the awareness of self. In his discussion of the evolution of language out of the sociality of primate grooming, Dunbar offers a discussion of the adaptiveness of being able to "mind read" others, what has come to be called "theory of mind" (1996, 85–105). "Having a Theory of Mind," he argues, "means being able to understand what another individual is thinking, to ascribe beliefs, desires, fears, and hopes to someone else, and to believe that they really do experience these feelings as mental states. We can conceive of a kind of natural hierarchy: you can have a mental state (a belief about something) and I can have a mental state about your mental state (a belief about a belief). If your mental state is a belief about my mental state, then we can say that 'I believe that you believe that I believe something to be the case.' These are now usually referred to as orders of 'intensionality'" (83). Dunbar notes that "intensionality" is a technical term, sometimes (but not always, as we see in Daniel Dennett's use of the term in chapter 5) used to distinguish it from "conventional intentions" (83).[7] This ability to reasonably anticipate the motives and actions of others, he argues, is a result of the significant neocortical development in primates and others I have already cited. Dunbar even suggests that "no living species will ever aspire to produce literature as we have it. This is not simply because no other species has a language capacity that would enable it to do this, but because no other species has a sufficiently well-developed theory of mind to be able to explore the mental worlds of others" (102).

Steven Johnson has surveyed experimental work in cognitive psychology that demonstrates empathetic understanding in children as young as four. "Human beings are innate mind readers," he writes:

> Our skill at imagining other people's mental states ranks up there with our knack for language and our opposable thumbs. It comes so naturally to us and has engendered so many corollary effects that it's hard for us to think of it as a special skill at all. And yet most animals lack the mind-reading skills of a four-year-old child. We come into the world with a genetic aptitude for building "theories of other minds" and adjusting those theories on the fly, in response to various forms of social feedback. (2001, 196)[8]

Johnson argues that our very sense of self-awareness—our sense of self altogether—is a function of the social-communicative skills of mind reading, skills of empathetic understanding. "Only when we begin to speculate on the mental life of others," he writes, "do we discover that we have a mental life ourselves" (196). In chapter 5, where I focus on the relationship of the brute physiology of pain in relation to the delicate semiotics of religious understanding and feeling, the crucial category is "theories of other minds," which give rise to supernatural religion and can be understood in terms of physiology, biological adaptation, and personal and social semiotics.

Here, though, I focus on hands. The intertwinings of hands and pronouns can be unpacked (or at least elaborated) in relation to Peirce's account of indexicality. After all, both pronoun and hand are quintessentially indexes. Moreover, it might well be said that cognition more generally is a species of index insofar as cognition, as Capra argues, is the activity of *specifying,* what I have been calling "attention" (and also "notation"). "According to [Humberto] Maturana," Capra writes, "perception and, more generally, cognition, do not *represent* an external reality, but rather *specify* one through the nervous system's process of circular organization. . . . [Maturana writes] 'Living systems are cognitive systems, and living as a process is a process of cognition'" (1997, 96–97). Moreover, there seems to be something inherent or "natural" to pointing. As Susan Goldin-Meadow notes, "Even before young children produce pointing gestures to orient another's attention toward an object, they respond to others' pointing gestures by directing their attention to the object indicated by the point (Lempers, Flavell, and Flavell 1977; Leung and Rheingod 1981; Murphy and Messer 1977)—unlike our dog, Kugel, who insists on looking at the end of my finger when I point out a toy to her" (2003, 91–92; Goldin-Meadow's references are included in the bibliography). In this example, young children "read" the gesture by reading other people's minds, the intension of the gesture. Goldin-Meadow calls gesturing "a natural accompaniment to speech" (144).

Here is Peirce's account of the "specification" of the index:

> No matter of fact can be stated without the use of some sign serving as an index. If A says to B, "there is a fire," B will ask, "Where?" Thereupon A is forced to resort to an index, even if he only means somewhere in the real universe, past and future. Otherwise, he has only said that there is such an idea as fire, which would give no information, since unless it were known already, the word "fire" would be unintelligible. If A points his finger to the fire, his finger is dynamically connected with the fire, as much as if a self-acting fire-alarm had directly turned it in that direction. . . . If A's reply is, "Within a thousand yards of here," the word "here" is an index; for it has precisely the same force as if he had pointed energetically to the ground between him and B. (1931–35, 2:305)

In this account, Peirce offers gestures that situate communication and, I suggest, attention and consciousness. Moreover, it situates communication and attention in relation to action, which is to say, in relation to narrative. Thus, Anne Freadman notes that Peirce's index accomplishes "the relation of knowledge and action: made, taken, or simply used as given, indices are devices for interpreting a bit of the world in such a way as to guide action . . . : some are devices for establishing the conditions of communication, others for using communication to some particular effect, and still others for converting 'experience' into the data for theory" (2004, 115).

The relationship between ideas and communication, word and touch, is directly linked to the connections between motor and linguistic functions examined in chapter 3. But this relationship can be seen in the very development of language. For a child, Wilson argues, the "compulsive, chaotic sounding-mapping of the world into the primary tool of her emergent powers of discrimination and intellect" is affected by "the progressive commingling of speech with the whole constellation of objects, people, and real-life situations encountered in the child's life" (1999, 193):

> During this stage when the thought–language nexus is building . . . something *very* important is happening in the hand itself. At about

> the age of one year the child's hands are rapidly becoming manipulative organs with fingers that will soon be able to move independently. The world of objects, and knowledge of the action of those objects, will increase rapidly, and distinctive actions which can be taken with objects in the hand will also increase. In other words, the thought–language nexus is becoming a hand–thought–language nexus. The child learns with real objects, by trial and error, to make constructions that are inevitably composed as discrete events unified through a sequence of actions. Playing with anything to make something is always paralleled in cognition by the creation of a story. Front→middle→back; beginning→middle→end; Steven→store→record.
>
> The brain is not a passive witness to the expansion of these mechanical and sensorimotor explorations and accomplishment. . . . [I]t models its processes and its formulations of the world on the narrative principle. (195)

The "narrative principle" Wilson describes is tightly aligned with indexicality: Peirce's index provides a starting point for events so that events become sequence. It is no accident that Vladimir Propp's catalog of narrative "functions"—that is to say, events necessary to folktale (or wondertale) narratives—include "departure," "spatial translocation," "return." (Greimas thinks so as well: see Schleifer 1987, 121–27.) "Here" and "there" are part of the logic or grammar of narrative. Moreover, *shared attention*—shared both in terms of storyteller and story listener (Greimas added the binary opposition of sender–receiver to Propp's single category of "dispatcher" [1983, 204]) and in terms of a narrative's gathering events into a sequence that can be grasped as a whole—is part of the logic of narrative.[9] Patrick Hogan has written persuasively about the "universal" connection between emotion and narrative—between the "iconic" nature of affect and the "indexical" nature of narrative. He argues that "biologically given proto-emotions—proto-fear, proto-anger, proto-affection, and so on— . . . are specified by socially functional practices and ideas [which themselves] . . . are to a great extent universal as well" (2003, 263). These social universals are organized and articulated in narrative. And the narrative articulation of sociality, as Wilson notes, begins with *indexical* engagements with the world.

The "revelation" of language that Helen Keller achieved with her teacher, Annie Sullivan, offers a dramatic depiction of this narrative phenomenon. Merlin Donald follows her story in his study of the evolution of consciousness: "One day, after many unsuccessful attempts to teach Helen words, Annie combined the spelling sequence for the word 'water' . . . with the act of immediately plunging Helen's hand into water. Although they had rehearsed this word often, without her realizing what Annie intended, this time Helen stopped for a moment and suddenly understood the connection between the hand signs and the meaning" (2001, 245). Helen's hand, the vehicle for both signs and sensations—vehicle for both symbol and icon—literally *embodies* Peirce's index in the action of engagement with the world. This dramatic narrative tells the story of the creation of what Wilson calls a "dictionary of manual semantics" (1999, 125), so that Helen's hand is transformed "into an articulate organ of expression" (145). "The brain," Wilson writes, "keeps giving the hand new things to do and new ways of doing what it already knows how to do. In turn, the hand affords the brain new ways of approaching old tasks and the possibility of undertaking and mastering new tasks" (146). "The hand," he concludes, "is involved from the beginning in the baby's construction of visuomotor, kinesthetic, and haptic representations of the world and the objects in it. This is a profoundly important role for the hand, whose importance in both cognitive and emotional ontogeny cannot be overstated" (197).

In the story of Helen Keller, Donald describes the remarkable connection between language and narrative, the ways in which language functions *first of all* as a bonding mechanism within a group, between parent and child,[10] between student and teacher: "The first priority was not to speak, use words, or develop grammars. It was to bond as a group, to learn to share attention and set up the social patterns that would sustain such sharing and bonding in the species" (2001, 253). This, of course, is close to Dunbar's description of the evolution of language as a replacement of primate grooming that allows members of a group to create bonds with a larger number of others than the time-consuming one-on-one process of grooming (1996, 78). Above all, Helen Keller *needed* a teacher. "Helen had grasped the general principle of connecting a sign with a meaning. Moreover, . . . the awareness of someone else's intention to communicate had kicked in, and the result was a major step toward acquiring language" (Donald 2001, 246). Donald continues:

> Despite the fact that she had so much rich cognitive material and all the apparent ingredients needed for language, including the raw objects of narrative description, event perceptions, agents, actions, situations, and models of three-dimensional space, she never used any

> of these to construct descriptions. Her isolated mind could not invent language as an internal means of thinking and representation, even under conditions where it would have been extremely important to do so. (248)

As Donald suggests here, above all "description" is *narrative description*—note his catalog of agents, actions, situations, as well as his global notation of "narrative description"—and such narratives, as Wilson suggests, are tightly connected—*homologically* connected—to the functioning of our hands. Even Peirce's indexical statement, "there is a fire," confuses the spatial-adverbial "there"—the *index*—of place with narrative's general "once upon a time there was as fire." Thus, Donald argues, "on a cultural level, language is not about inventing words. It is about telling stories in groups. *Languages are invented on the level of narrative, by collectivities of conscious intellects*" (292). Peirce's indexical gesture—like the overwhelming hand-plunging gesture of Helen Keller, forced upon her by another person—situates meaning in a world in which "agents, actions, [and] situations" combine into the interpersonal relationships of stories *located* in the world where things happen. As Freadman says, communication is made possible, it creates effects in the world, and it makes possible reflection about communication itself (see 2004, 115).

Indexical Handedness

The connections between linguistic and motor functions—between ideas and communication, words and touch—are bound together in Peirce's index. These connections have been studied in relation to human adaptations. Chris McManus argues that "language is not the only task requiring precision timing; another such is throwing" (2002, 224), and Dunbar notes that "aimed throwing is clearly important for hunting, so one obvious conclusion is that language evolved on the back of throwing. The fine motor control needed for aimed throwing, so the argument runs, provided us with the neural machinery for fine motor control of the organs of speech" (1996, 134). (These claims nicely build on Lieberman's examination of the motor aspects of language.) McManus further argues that these connections are closely related to the development of asymmetric handedness. Although Dunbar is not altogether convinced, McManus offers a remarkable analysis of human evolution, focusing on "what happened two or three million years ago that made the right and left hemispheres [of the brain] different, resulting in the development of right-handedness and language" (2002, 225).[11] What happened, he speculates, is that brain lateralization into to asymmetric hemispheres was directly related to the asymmetry of the heart on the left side of the body of vertebrates and that the "speeding up" of the development of the left hemisphere of the brain facilitated "spoken language, grammar and precision control of tools" (227). Discussing handedness, Wilson also makes an observation that might help us understand the ways in which lateralization accommodates language, especially in relation to Peirce's situating of the index between iconic and symbolic representations.

Peirce opposes the facticity or external aspect of the *index*—its referential indication of something *in the world* (such as smoke indicating fire)—with the different internal aspects of *icons* and *symbols:* against the facticity of the *index, iconic* representations depend on the phenomenological similarities of sensate experiences and offer something similar to what is being represented (such as Washington's image on a dollar bill); and *symbolic* representations offer conceptual and conventional signs—*arbitrary* signs—for what is being represented (such as the word/sound "tree" to represent a certain plant). All three aspects of signs, with their internal and external aspects, function together. Thus, the repeated similes in Crane's poem comparing the knots and notches in the worker's hand to "the marks of wild ponies' play" combine similarity, causal connection (in the violence that produced the marking of hand and turf), and the accidental arbitrariness of the fact that "ponies'" and "play" both begin with /p/. Susan Goldin-Meadow also finds these three levels of signaling in hand gestures: she describes gestural "emblems"—for example, the thumbs-up—that can occur without any combination with speech (2003, 16) and three classes of gestures that accompany speech. The first, like emblems, are "motor" or "beat" gestures that repeat and emphasize speech patterns; the second are "deictics" that, like Peirce's index, point to things; and the third are "lexical gestures" that "convey meaning [that] is fundamentally different from the method used by speech" (6, 23). Lexical gestures "express knowledge in gesture that [the communicators] do not express in speech" (57), knowledge that "is neither fully explicit nor fully implicit" (60).

"In writing," Wilson writes, "as in stone tool manufacture, the dominant hand's performance is micrometric, rehearsed, and for the most part *internally* driven. The performance of the nondominant hand is macrometric, improvisational, and *externally* driven" (1999, 162). The rehearsed and internally driven functioning of the dominant hand—based, as both McManus and Wilson note, on the lateralization of the brain and brain functions—is closely related to the work of symbolic discourse, by means of which, as Peirce says, "such an idea as fire . . . would give no information, since unless it were known already, the word 'fire' would be unintelligible" (1931–35, 2:305). The work of the nondominant hand—and of the right brain more generally—is, like the index and like evolution,[12] externally driven: it is work in the world. "We know now," Dunbar asserts, "that the right hemisphere is specialized for the processing of emotional information. There is evidence to show that emotional cues are detected faster when they are on the left side of the visual field (transmitted across to the right side of the brain) than on the right side" (1996, 136; see also Trimble 2007, chapter 4). Such information, he shows, is cued by the environment: sexual rivals, babies, predators. More generally, it is cued by a wider sense of context and the comprehension of wholeness, Greimas's vague sense of "meaningful whole"—especially the meaningful whole narrative gives rise to. Thus, Robert Ornstein notes that "people whose left-hemisphere damage has left them with aphasia can use gestures, facial expressions, or tone of voice, grunts, etc., to request what they need. And when they are asked to do something, they respond in context. . . . [They] had lost function of the left hemisphere and thus lost the text, but not the context, of . . . communication" (1997, 98–99).[13] Ornstein continues, "The right hemisphere decodes the external information that we use to compute context; it helps assemble the whole field of view to create an overall understanding of a scene. It helps us understand where we are figuratively, in terms of who we think we are, and literally, in space, in time, and in our field of vision" (101). In a specific example, Ornstein notes that "right-hemisphere-damaged subjects," viewing a film of a scene of two people sitting in a living room who are asked "Can you play tennis?" think the "scene is OK in which the listener responds by playing tennis, right there, in the living room." He concludes that "people with right-hemisphere damage can always understand the literal meaning of a request, *but they cannot always judge what the request means in context*" (103). McManus notes that the differences between left and right brain functioning, "in the final analysis, . . . must be due to some low-level difference associated with neurones and their connections. Differences of that sort might eventually make one hemisphere [the left brain] better for the serial processing of symbols and hence for analytic and logical high-level analysis, and the other [right-brain hemisphere] best suited for the parallel processing of images" (2002, 222–23). The dominant right-handedness of humans (about 90 percent of people) is created by the dominance of the left brain, which is also associated with language.

An important corollary to this evolutionary history is the question of why 10 percent of the population remains left-handed. McManus argues that the gene determining left-handedness possesses the

> ability to confer randomness on the organisation of the brain, not only for manual dexterity and language . . . , but almost certainly for a host of other cerebral asymmetries, such as those for reading, writing, visuo-spatial processing and emotion. Although it might seem paradoxical, randomness, at least in small amounts, can benefit complex systems. The idea I will present here, which I call the *theory of random cerebral variation,* suggests that human brains are not all alike, some having fundamentally different organisations, and that such variation sometimes produces brains better able to carry out particular complex tasks. (229)

This "theory," already mentioned in chapter 2, offers an argument for the level of semiotic organization of disorder growing out of order while it suggests that Peirce's indexical meanings are at least somewhat connected to the functioning of the right brain. It can also help us to understand the power of Crane's poem in which the confusion of pronominal antecedents and the work of the poem's similes to gather and cluster meanings—including the poem's governing confusion of the literal and figurative significances of the hands themselves—support the overall effect of the poem, what Greimas might call its "meaningful narrative whole."

McManus offers an argument for "the right hemisphere's knowledge of hidden meanings." "As well as being involved in understanding three-dimensional space, having knowledge about the world, and being involved in humour," he argues,

> the right hemisphere has other functions as well. It is particularly involved in attention, a process central to cognition yet often ignored, forgotten or unnoticed. Without the ability to focus attention, the mind would be flooded with irrelevant, pointless, useless information that would overwhelm it. (2002, 185)

Such attention—as well as "having knowledge about the world"—is, as we have seen, the work of Peirce's index: to call attention to things of the world. Attention is at the heart of Donald's discussion of "level-1 awareness," the ability of organisms—bees as well as humans—to grasp events from the welter of stimuli, the "capacity to perceive objects and abstract events" (2001, 183) that he calls "binding."[14] In any case, McManus suggests that some kind of complex binding of phenomena to achieve apprehension might well be a function of lateralization in the human brain. "The left hemisphere," he notes, "knows how to handle logic and the right hemisphere knows about the world. Put the two together and one gets a powerful thinking machine" (2002, 184; see also Ornstein 1997, esp. ch. 8).

Moreover, putting them together creates what I might call a powerful linguistic machine. Ornstein notes that

> both halves of the brain are needed for the two elements of everyday language. The left side looks after the basic text, the conventional features of language: choice of words, syntax, and literal meaning. But taking part in a conversation requires a lot more than using the right words in the right order and knowing what individual sentences mean. To understand fully what someone is saying, you have to be able to interpret his or her tone of voice, apply the conventions of polite conversation, follow a narrative, understand gestures, and so on. You need to know when sentences don't have their usual function and be able to fathom the speaker's purpose. (1997, 113)

Ornstein's chief example of "following a narrative," as it is Greimas's example of "grasping" the meaningful whole in *Structural Semantics*, is the joke (106; Pascal Boyer also notes the "salience" of jokes [2001, 133]), and he emphasizes the material nature of this phenomenon:

> The right side [of the brain] doesn't (mystically) somehow perceive things whole, but seems to be specialized for the large elements of perception, the overall shapes of objects, the word shape, the information contained in the size, sounds, and intonation of words strung together. These convey emphases and much of the subtext and contextual meaning. It handles the large movements of the limbs, and the larger emotional reactions such as anger and disgust. And the left side handles the small, precise links that carry the smaller, more precise meanings and movements. It's this specialization that contributes to one side being good for the analysis of the small elements versus the

> synthesis or holistic vision, or language via the literal meaning versus the intonation and indirect meaning. (1997, 175)

Grasping the meaningful whole is beyond a mechanical operation: it gathers up the sense of things *from* the details of a visual array[15] or the particular words of a sentence or sentences in a paragraph or story. (Ornstein notes that right-hemisphere processing allows us to see how "sentences . . . relate to one another" [117], which is another narrative function.) All this suggests that while the left brain organizes or processes the logic of Peirce's symbols, the right brain seems to be related to the worldliness of Peirce's index.

Articulations of Language

The two operations McManus and Ornstein describe are systematically repeated in the relationship between the two "articulations" of language that linguistics has isolated, that of meaning/semantics and that of sounds (or "sound-images")/phonology.[16] Moreover, this relationship is closely linked to the evolution of language and, indeed, the evolution of consciousness. The linguist André Martinet first coined the term "double articulation" to describe this defining feature of human language, that language constructs and organizes meaning (and meaningful distinctions) and also constructs and organizes the communicative system (the *combinable* system of sound or sound-images or of the systematic array of gestures in sign languages).[17] He argues—as do Mithen and Donald working in the far different discipline of evolutionary anthropology—that the double articulation distinguishes language from other signs (such as various primate calls). His work suggests that the systematic double articulation of language describes the relationship between Peirce's index and symbol in a way that also suggests the constitutive nature of gesture altogether. That is, the first articulation of meaning is more or less symbolic insofar as it is based on its own arbitrary logic of distinction and relationship ("the law that will govern the future," as Peirce puts it [1931–35, 1:23]), while the second articulation is more or less indexical insofar as it deals with worldly material objects, sounds, gestures, and orthographic marks. The work of the first articulation of meaning, Martinet argues, are "additive," while those of the second articulation of phonemes are "destructive." The presence of a phoneme, the material sound pattern of language, he writes,

> signalizes that any inference, as to the meaning of the utterance, which might be drawn from the context considered without it is wrong: if, to the statement *it is good,* I add *very,* I am just adding some additional information without deleting what was previously there, but if to the statement *it is a roe* I add a /d/ phoneme, the statement becomes *it is a road;* one element of information *roe* is deleted and replaced by another one. (1962, 36)

It is the function of Peirce's index to delete and replace: the dynamic connection of pointing the finger deletes the "here" of communication to the "there" of the fire. Index inhabits the level of the second articulation of the sound-images of phonemes, as Saussure calls them (while icon inhabits the level of sensate sound itself); the second articulation is simple in comparison to the first articulation of meaning.[18]

Moreover, as Martinet suggests, the systematic relation between these articulations (or "duality of patterning") is crucial to language itself. Michael Corballis, in his study of the evolution of language, *From Hand to Mouth,* describes the particular problem of the systematicity of language.

> Although duality of patterning is considered a hallmark of true language, it has also been described as "one of the most difficult problems of language evolution." How did we evolve a system that is structured in such different ways at two entirely different levels, one involving sound and the other symbols? I shall suggest . . . that the mystery is largely solved if language evolved from gestures, and not from vocalizations. (2002, 114–15)[19]

Later, he cites the argument of the neurologist Giacomo Rizzolatti that "the importance of Broca's area in the evolution of speech is not that it was involved in movements of the mouth and face but rather that it served to map internally generated hand movements onto perceived hand movements made by others" (152). More generally, Corballis claims that "the emergence of articulate speech" would appear "somewhat less miraculous if we suppose that it was built on a scaffolding of manual communication that had deep roots in primate evolution" (155–56).

Among the scholars Corballis cites is Merlin Donald, whose *Origins of the Modern Mind* argues for the emergence of language from gesture in the process of "cognitive evolution." Specifically, Donald suggests that hominids developed a "mimetic culture" of organized gestures that evolved into a linguistically organized "mythic" culture and subsequently evolved, more rapidly (since it was based on inventions of writing), into a literately organized "symbolic" culture. "Mimetic skill or mimesis," he argues, "rests on the ability to produce conscious, self-initiated, representational acts that are intentional but not linguistic" (1991, 168). (In this, it is significantly different from the "second articulation" of Martinet, which is linguistic precisely because it is *systematically* related to language's "first articulation.")[20] Mimesis, Donald argues, "is fundamentally different from imitation and mimicry in that it involves the *invention* of intentional representations" (169). "Mythic thought, in our terms," he says, "might be regarded as a unified, collectively held belief system of explanatory and regulatory metaphors. The mind had expanded its reach beyond the episodic perception of events, beyond the mimetic reconstruction of episodes, to a comprehensive modeling of the entire human universe. Causal explanation, prediction, control—myth constitutes an attempt at all three, and every aspect of life is permeated by myth" (214). Here is a chart suggesting the parallels among science, semiotics, linguistics, and evolution:

Schrödinger [science]	**Peirce** [semiotics]	**Martinet** [linguistics]	**Donald** [cognitive evolution]
physics (order from disorder)	icon	[noise]	mimicry (episodic perception)
biology (order from order)	index	second articulation (sound pattern; phoneme)	mimesis (patterned gesture)
semiotics (disorder from order)	symbol	first articulation (meaning; semantics)	myth/symbol (language)

To this list I could also add the resources of language Crane calls upon in his poem: the sounds ("noise"?) of his words, the "knots and notches" and "ponies' play," are sensate experience; the similes, "like wings of butterflies / Flickering in sunlight over summer fields" or the notches that "were like the marks of wild ponies' play," function and gesture like indexes; and the meaningful whole of the poem as a whole—where gestures of hands and eyes, similes of nature, and the caretaking activity of the whole scene—creates a sense of "warm friendship" growing out of the poem's narrative.

Skills of the Hand

The mimetic invention of unsystematic intentional representations Donald describes—like particular acts of pointing and attention—confuses the distinction between the second and first articulations of language insofar as such inventions are repeatable but not systematically combinable. It also confuses the distinction between the material world and the self. The gestures of mimesis, Donald suggests, are "characterized by the integrated use of hand and limb movements, along with a combination of facial expression, sound, and gesticulation" (1991, 178; he is talking about primates in general). But he is particularly interested in toolmaking, since it was, he says, "probably the first instance of behavior that depended entirely upon the existence of self-cued mimetic skill. The reproduction [of tools] could not be dependent on immediate environmental reinforcers or contingencies" (179; see also Wilson 1998, 33). Here, Donald is describing the truly amazing fact that tools—flint knives, stone blades, even musical instruments—were made *in anticipation of their use* (and thus participating in "the law that will govern the future" Peirce describes). It is skills of the hand, Donald suggests, that, adaptive in themselves, also allowed the adaptations of communication skills—including the "tool" of the second articulation—and the confusions of the negative sciences, as I have described them, of semiotics and information. That is, the possibility of confusing a sentence with a noun requires the same abstraction (or "displacement") from "environmental reinforcers" as do both the double articulation of language and cognition more generally. (For a powerful discussion of the adaptation of the hand to toolmaking, see Napier 1993.)

The adaptation of the human hand is remarkable. In his wide-ranging study, Frank Wilson describes three perspectives in studying the hand: mechanical, biological (evolutionary), and neuro-cultural (1998, 10). (In this, Wilson is repeating the system of hierarchical levels with which I am working: physics, biology, semiotics. That he hyphenates "neuro-cultural" is an instance of the phenomenon at the heart of Tourette syndrome.) He offers wide examples of the last category—toolmaking, gesticulating, even the motor skills of juggling—but the most striking (he begins his book with this) is the example of instrumental music as a category of communication that is "neither gesture nor sign." "Musical skill," he writes,

> provides the clearest example and the cleanest proof of the existence of a whole class of self-defined, personally distinctive motor skills with an extended training and experience base, strong ties to the individual's emotional and cognitive development, strong communicative intent, and very high performance standards. . . . [Instrumental] musical skill [provides] evidence of the existence of a whole *family* of physically and cognitively demanding, hierarchically structured, creatively rich human skills that (like sign) have communicative content and are "put out through the hand." (207)[21]

Such skills are indexical in the literal sense—they are finger skills—but they are also Peirce's index insofar as they situate communication and are, as Freadman says, "devices for establishing the conditions of communication, [or] . . . for using communication to some particular effect, [or] . . . for converting 'experience' into the data for theory" (2004, 115). If this is so, then it might well be that an important element of music (as Mithen describes it in his speculative evolutionary history) is its indexical nature.

One striking feature of the skills Wilson describes is that they can be performed as intentional actions that are nevertheless "unconscious." Following the work of Ernest Hilgard, Ariel Glucklich notes that the phenomenal sense of a "totally unified consciousness" is a functional subsystem that is "functional and not ontological." Although subsystems "can interact, many functions take place independently, and often without the awareness of the 'central' consciousness. Long-distance car driving can exemplify the relative autonomy of motor-control operations, as do typing, playing the piano, or signing a check" (2001, 119). Such subsystems, as I note in chapter 5, are *functional* in the sense that they are defined, as evolutionary adaptation is, in relation to their goals rather than their physical ("ontological") existence. They are, as Glucklich says, "functional rather than anatomical" (91). Glucklich's examples include "the circulatory system, the reproductive system, the immune system, the nervous system, and so forth," and he argues they are not anatomical because "the same organs can be 'shared' by more than one system" (91). Other "subsystems" he could have mentioned are those of hand gestures, the minute work of fingers, the system of throwing, and, as he says later, the motor-control operation of playing the piano. Such subsystems—again, like evolutionary adaptation—are *material* even if they are not essentially "physicalistic." They are determined by the relationships among material properties: they are a function of the "single cells, then individual organs, and moving up the scale there are complex functional systems" that Glucklich describes (91), and also a function of the material environment in which the organism and community exists. Edelman's global brain theory of "neural Darwinism" describes such subsystems.

Narrative Contingencies

Finally, like the "goal" orientation of evolutionary adaptation, such subsystems follow the lead of the indexical action of hands by being essentially narrative in the organization, the creation of order out of order. This is the great insight of Stephen Jay Gould in his studies of natural, evolutionary history. Such history, he argues, is "irreducible" (1989, 277) and also essentially "contingent" (278). "Historical explanations," he writes,

> are distinct from conventional experimental results in many ways. The issue of verification by repetition does not arise because we are trying to account for uniqueness of detail that cannot, both by laws of probability and time's arrow of irreversibility, occur together again. We do not attempt to interpret the complex events of narrative by reducing them to simple consequences of natural law; historical events do not, of course, violate any general principles of matter and motion, but their occurrence lies in a realm of contingent detail. . . . And the issue of prediction, a central ingredient in the stereotype [of "scientific method"], does not enter into a historical narrative. We can explain an event after it occurs, but contingency precludes its repetition. (278)

He argues that "history presents two special problems: (1) frequent absence of evidence, given imperfections of preservation; and (2) uniqueness of sequences, unrepeatable in their contingent complexity, and thereby distancing the data of history from such standard concepts as prediction, and experimentation" (102).

Gould describes, in explaining Darwin's practice, the "homological" reasoning I have referred to throughout these chapters, reasoning from "the material reality of history" that he claims for Darwin (1986, 68). Such reasoning attempts what Paul Ricoeur describes in a more abstract philosophical language as a particular modality of explanation, the "synthesis of the heterogeneous" that organizes a series of events "into an intelligible [meaningful] whole, of a sort such that we can always ask what is the 'thought' of this story" (Ricoeur 1984, ix, 65). Specifically, Ricoeur situates this modality in relation to two other modalities of explanation, the "theoretical mode" that is "represented by Laplace's system" of mechanical reduction and the "categoreal mode" that aims at determining "what type of object we are dealing with, what system of a priori concepts organizes an experience" that is represented by Plato and systematic philosophy more generally. The synthesis of the heterogeneous is the "configurational mode" that "puts its elements into a single, concrete complex of relations" and that "characterizes the narrative operation" (159; see also Schleifer 2000a: 13–16). Such a complex of relations, Gould argues in examining Darwin's historical science, organizes itself around recognizing similarities, whether they be homologous or analogous: "Historical study," he writes, "manifests its special character by placing primary emphasis upon comparison and degrees of similarity, rather than the canonical methods of simplification, manipulation, controlled experiment, and prediction" (2002, 102). Still, Gould's emphasis on contingency supplements Darwin's creation of biological order out of the preexisting order of the physical world, his sense of the "uniformity" of "gradual adaptive change within local populations" (1980, 15) that, Gould says, was Darwin's "finest achievement," since "for the first time, [it] made evolution a workable research program, not just an absorbing subject for speculation" (1986, 61; see also 2002, 105–06). Such contingency supplements biological science, Gould says, by requiring "sequencing" (2002, 106–8), or what he earlier called "inferring history from its results" (1986, 62), and also requiring "consilience" (2002, 108–10; 1986, 65–66).

Such sequencing is a version of Peirce's abduction, the inference of a *narrative* sequence from fragmentary data often by means of focusing on the "form of discordance" of "unique objects" such as the giraffe or the panda's "false thumb," which notices (again, like Peirce) "some imperfection or failure of coordination between an organism and its current circumstances" (Gould 2002, 104; see also 1986, 64–66). Such imperfections quite often are noticeable precisely because they are apprehended, as Peirce suggests, in the context of different *kinds* of fact. In this chapter on hands, I should say something more about Gould's repeated reference to the "the highly inefficient, but serviceable, false thumb of the panda," which he describes in

"Evolution and the Triumph of Homology" more generally as the "principle of imperfection." This "thumb" was "fashioned from the wrist's radial sesamoid bone because the true anatomical first digit had irrevocably evolved . . . to limited motility in running and clawing" (1986, 63). Gould alludes here (and in *The Structure of Evolutionary Theory,* 2002, 104) to his earlier book, *The Panda's Thumb,* where he notes that the "false" opposable thumb "comes equipped not only with a bone to give it strength but also with muscles to sustain its agility" (1980, 22) and argues that "the proof of evolution lies in imperfections that reveal history" (13). Such imperfections are "odd arrangements and funny solutions . . . —paths that a sensible God would never tread but that a natural process, constrained by history, follows perforce" (13, 20–21). More generally, he argues that "if organisms have a history, then ancestral stages should leave *remnants* behind. Remnants of the past that don't make sense in present terms—the useless, the odd, the peculiar—are signs of history" (28). They are also signs of what John Casti calls the "complexity" of complex systems that, necessarily, "is a joint property of the system *and* its interaction with another system, most often an observer and/or controller" (1995, 269). The intersection of two systems, as Peirce suggests, is the object of analysis in abduction, which Max Fisch has called his "major single discovery" (1981, 17).

But more than sequencing and consilience, I think, Gould emphasizes contingency—both in his remarkable case history of the Burgess Shale in *Wonderful Life* (1989) and in his larger theoretical examination of evolution in *The Structure of Evolutionary Theory* (2002)—by arguing that the material evidence of long biological history suggests that Darwin's uniformitarian argument be supplemented with a sense of the "punctuated equilibrium" of evolution. In fact, he argues that "gradualistic anagenesis [progressive evolution] occurs only rarely in nature" (2002, 606) and that punctuated equilibrium "holds that the great majority of species, as evidenced by their anatomical and geographical histories in the fossil record, originate in geological moments (punctuations) and then persist in stasis throughout their long durations" (766).[22] The most dramatic instances are those of mass extinction, such as the record of the Burgess Shale Gould examines in *Wonderful Life,* which "seem to be genuine disruptions in geological flow, not merely the high points of a continuity. They may result from environmental change at such a rate, and with so drastic a result, that organisms cannot adjust by the usual forces of natural selection. Thus, mass extinctions can derail, undo, and reorient whatever might be accumulating during the 'normal' times between" (1989, 305). But more generally—and perhaps, as Gould says, in "an overarching hypothesis"—"punctuation records something quite general about *the* nature of change itself" so that "the differing causes of punctuational change at each level—the waiting time between favorable mutations in bacterial anagenesis . . . , the scaling of ordinary speciation as a geological moment in punctuated equilibrium, or the simultaneity of species death in mass extinction—must run in a common structural channel that sets and constrains the episodic nature of alternation" (2002, 929). These levels of mutation, "momentary" speciation, and mass extinction—possibly analogous to physiology, evolution, and the "negative science" of semiotics as well as to Ricoeur's three modalities of explanation, Laplace's mechanism, Plato's speciation, and the "configurational mode" of narrative comprehension—are each marked by the irruption of disorder into order.

Gould is careful to maintain that the punctuated equilibrium must be understood in terms of the "material reality" of homological connections within the fossil record—it must be understood in term of macroevolution of geological time—but he still notes that "many realities of late 20th century life (from the juggernaut of the Internet's spread to the surprising, almost sudden collapse of communism in the Soviet Union, largely from within) . . . raise the general critique of gradualism" (2002, 922). Within science, he notes, twentieth-century mathematics seems to "share a common intent to formalize the pattern of small and continuous inputs, long resisted or accommodated by minimal alternation, but eventually engendering rapid breaks, flips, splits or excursions in systems under study: in other words, a punctuational style of change" (922). John Casti studies such mathematics in *Complexification,* where he argues that complexity arises in the "interaction" of different systems (1993, 269). Such interaction is configurational, and it situates itself in relation to Ricoeur's modalities of explanation (and also to Casti's description of "the well-tested principles and laws of classical physics," the "semi-physics" of developmental biology, and the "metaphysics" of catastrophe theory [1993, 77–80] I examined in chapter 2). In a way, both Ricoeur and Casti describe the creation of order out of disorder, order out of order, and—if we emphasize the *contingency* of narrative explanation the way Gould does—disorder out of order. The referentiality of indexicality—its power of focusing attention on an already ordered state of affairs—makes it seem a species of categoreal understanding (after all, smoke does indicate fire). Yet it also accomplishes, as Freadman notes, "the relation of knowledge and action" (2004, 115) in its gestures of narrative.

Gestures of Narrative

In this context, we can isolate, perhaps, the indexical aspect of poetry. Take Keats's late poem, "This Living Hand":

> This living hand, now warm and capable
> Of earnest grasping, would, if it were cold,
> And in the icy silence of the tomb,
> So haunt thy days and chill thy dreaming nights
> That thou wouldst wish thine own heart dry of blood
> So in my veins red life might stream again,
> And thou be conscience-calmed—see here it is—
> I hold it towards you. (1967, 1205)

In this poem, Keats "plays" his hand, so to speak, transforming its mechanical–biological nature—its ability to grasp and touch—into a ghostly symbol of remorse articulated through vampiric anger. It accomplishes this symbolic achievement by means of the contrary-to-fact subjunctive, the very grammatical category that will govern the future: Keats's hand would, if it were cold and in the icy tomb, so haunt Fanny Brawne's days and nights that she would wish to revivify Keats with her own life. George Steiner has argued that counterfactual conditionals and the future tense—precisely because they do not exist in the real world of "environmental reinforcers"—comprise the genius of language in which, he says, "we can *say anything*" (1975, 216). Greimas and Saussure, in a very different tradition from Steiner, would agree: the genius of language for them is that it complicates referentiality—it "has to do with reality," as Shoshana Felman has argued (2002, 76; see also Schleifer 1987, 201–6)—and involves itself as part of the material world. (Crane, as we have seen, elaborates his hands through similes rather than counterfactuals.) Yet there is nothing counterfactual about Keats's hand itself. It is, as Peirce says, "somewhere in the real universe, past and future," and it needs an index to situate it there. Such indexicality, in Keats's poem, breaks up the lines themselves: his hand, held before him, held out to Fanny, offers the contrary of life and death, the very systematic confusions of presence and absence (as different classes of fact)[23] that, based on index and hand, is the genius of symbol. Yet even while it breaks up the lines, it tells a counterfactual story. Keats himself (and his hand) evolved out of communities of hominids, struggling in the real world, using hand and voice, as Kenneth Burke has said of literature, as equipment for living (1994), Said's *worldly* work. Index becomes symbol—a vehicle for narrative—in the way that the iconic signal of automatic subcortical firings can also be taken up by poetry.

This might be clearer if I present a dead hand rather than a living one. Listen to this poem, "Meditation on Certainty," by Jacques Roubaud.

> The door pushed back the light.
>
> I knew there was a hand. who could from now on grant me the rest?
>
> Having seen, having recognized death, that it didn't just seem, but was, there was, certainly, no sense doubting it.
>
> Having seen, having recognized death.
>
> If somebody had said: "I don't know if this is a hand." I could not have replied. "look closer." no language game could budge this certainty. your hand hung down from the bed.
>
> Almost warm. almost. still almost warm.
>
> Blood coagulated at the fingertips. like dregs of Guinness in a glass.
>
> I couldn't see it as human. "there's blood in any human hand." I understood this proposition very clearly. because I was seeing it confirmed by its negative.
>
> I didn't have to tell myself: "blood flows through any living hand." though it's a thing no one has ever seen. the blood here obviously not flowing. I could not doubt it. I had no reason to. (1990, 11–12)

Roubaud is mourning the death of his young wife, in the presence of her dead body, and the major linguistic resource he has in the face of death, like Keats, is the index: the imperative *look closer.* Hand here, as in Keats, is life itself: not grasping now, not warm, but the occasion for self-reflective internal dialogue. Roubaud is speaking to himself, he is left alone with only himself; and *this* indexical situation, the situation of communication, is itself *situated* within the context of shared cognition, the context of semiotics. Moreover, it is emotionally powerful—"the feeling of what happens," as Damasio says—and its emotion arises simply by its making speech itself contrary to fact: if somebody had spoken, Roubaud says, I could not have replied.

In this poem—as in Keats—the hand itself, like Peirce's index, is occasion for story, recognition, and cognition: for dialogue, for situating communication in the world, for making language work. The hand's toolmaking, Peter Reynolds has argued, is a form of "heterotechnic cooperation" in which "each individual would anticipate what the other was about to do and facilitate it by performing the complementary action" (1993, 411): it calls for a "theory of mind" that, Dunbar notes, produces literature (1996, 102). That is, it is with the index that the real enters language: the constraints of adaptation, the adaptive rewards of cooperative gesture, the "real" of reference. Jacques Lacan has argued that "the real" (as opposed to the imaginary and symbolic in another tripartite system that could be superimposed on my chart),[24] "whatever upheaval we subject it to, is always in its place" (1972, 55), but indexicality brings the world of the real—matters of fact—within the purview of language and discourse; it subjects it to story, cognition, and recognition. This is clear in both Roubaud's and Keats's poems, and even suggested in Crane's: in Keats and Roubaud, death—which is only hinted at in Crane—is, after all, most real. "Just in case you thought there was no distinction between representation and reality," Regina Barreca has said, "there is death. Just in case you thought experience and the representation of experience melted into one another, death provides a structural principle separating the two. See the difference, death asks, see the way language and vision differ from the actual, the irrevocable, the real?" (1993, 174).

The actual, the irrevocable, the real are all subject to be indicated and recognized by Peirce's index and by the human hand itself, pointed to, pointed out. With this discussion of language, attention, and our hands, I am suggesting that the *recognition* of language—the recognition most pronounced in Keats's and Roubaud's narrative confrontations with death and with the real, but a recognition that can always take place in linguistic-narrative activity in relation to fire or caretaking or an interlocutor—is neither sign nor gesture. Rather, such recognition arises out of sign and gesture, in the presence of meaning (and even the negations that gather around and within meaning) and in the presence of the real (and even the negative reality of death and nonbeing, the darker sides of materialism I mentioned in chapter 2), and it arises as a manner in which we do work in the world.

5. Pain, Memory, and Religious Suffering: Materiality and the Subject of Poetry

> Pain always arrives with a hidden narrative. Science prefers the same thing to happen, in the same way, over and over, but pain is subjective, invisible, multi-faceted, and individual. This is why science and pain have been uneasy bedfellows, whereas the shifting valence of suffering is a central theme in literature.
>
> —*Marni Jackson,* Pain: The Fifth Vital Sign

> In the realm of suffering, affliction is something apart, specific, and irreducible. It is quite a different thing from simple suffering. It takes possession of the soul and marks it through and through with its own particular mark, the mark of slavery. . . .
>
> —*Simone Weil, "The Love of God and Affliction"*

In this chapter, I describe the relations among pain, memory, and religious suffering. In doing so, I take up, in another register, the physiology of voices I examined in Tourette syndrome, the communality and semiotics I examined in the evolution of the work of our hands, and the hierarchy of materialism I am tracing throughout *Intangible Materialism* altogether. This is clear in the particular nature of pain, which, as Ariel Glucklich says, "is conscious by definition" (2001, 96; see Jackson 2002, 18, 148) even while, as Roselyne Rey notes, the "anatomical and physiological foundation" of pain makes it in important ways a *material* and "not an historical subject in the same sense as fear, or hell, or purgatory"; pain, she says, is "certainly [the] one experience where the human condition's universality and the species' biological unity is manifest" (1993, 5). If Tourette syndrome is unconscious (or at least wholly unintentional) and the work of our hands is often intentional but not fully conscious, then pain—and especially its form comprehended under the categories of suffering and religious suffering—presents a remarkable instance of consciousness fully imbricated, as it is, in the materiality of its physiology and its adaptiveness. And *because* it is necessarily conscious, it is intimately tied up with any notion we might have of self or personhood, and tied, as well, with the senses of subjectivity that literature creates in its discourses. The experience of pain, and especially religious suffering and that other ghostly phenomenon, phantom pain, might well be a clear and defining moment of consciousness that can help us comprehend *its* materiality. Indeed, it is my hope that the three topics of this chapter—pain, memory, and suffering—themselves encompass the material physiology, biological adaptiveness, and semiotic explosions of meaning I examine throughout this book.

The Memory of Pain

It seems to me that we cannot examine memory, including the memory of pain, without also examining anticipation—which hints at that "law that will govern the future" Peirce talks about (1931–35, 1:23)—as both a physiological fact and a phenomenological experience. Patrick Wall, who (often along with Ronald Melzack) helped shape the study of pain in the twentieth century, nicely describes the physiological fact of anticipation in the chapter on "The Placebo Response" in his book *Pain: The Science of Suffering*. He cites a number of animal and human experiments in which the "expectation" of certain effects led to measurable physiological changes in the subjects. "If a rabbit has experienced a series of small insulin injections that decrease the blood glucose and is then given a saline injection in the same conditions, the animal reacts by raising its blood glucose. The animal has learned to counteract the effects of the drug by raising its blood sugar. With a saline injection, it reacts as though it has received the insulin" (1999, 160). In people, a parallel experiment demonstrated a similar "placebo response," but when participants were told that the stimulus was being manipulated (parallel to the saline replacing insulin in the case of rabbits), no placebo response was recorded. This result suggests to Wall that the placebo response is not simply a Pavlovian "conditioned response" (168) but also, on some level, a cognitive response. Moreover, it leads Wall to *define* pain in relation to anticipation: "Pain," he proposes, "occurs as the brain is analysing the situation [that gave rise to pain] in terms of actions that might be appropriate" (169; see also 177); "Pain," he says, "is then best seen as a need state, like hunger and thirst, which are terminated by a consummatory act" (183). That act, however, unlike hunger and thirst, is not directed toward objects in the world (food or drink) but rather focuses on states of the body: as a "proprioceptive" warning[1] or the imposition of rest.

Poets often agree with Wall's sense of the anticipatory nature of both memory and pain. When T. S. Eliot describes April as the cruellest month in *The Waste Land,* he notes that it mixes memory and desire, though many other poets—and perhaps Eliot himself—describe the mixture of memory and dread as well. Memory—and especially the memory of pain—creates predispositions, tendencies of response, ways of being in the world. Even the simple memory of conditioned response, as in Pavlov's dogs and Wall's rabbits, is always marked by anticipation or dread. Dread, however, might well be the more potent marker of memory; Robinson Jeffers has said this in another poem: "Happiness is important, but pain gives importance." In fact, Antonio Damasio has argued that "pain and pleasure are not twins or mirror images of each other, at least not as far as their roles in leveraging survival [in biological adaptation]. Somehow, more often than not, it is the pain-related signal that steers us away from impending trouble, both at the moment and in the anticipated future. . . . There seem to be far more varieties of negative than positive emotions" (1994, 267).

The negative "emotion" of pain—elsewhere Damasio argues that pain is not exactly an emotion (1999, 71; 2003, 32–34)—is literally and materially marked and inscribed in damaged tissue. Wall cites "a modern definition of pain": "Pain is an unpleasant sensory and emotional experience associated with actual or potential tissue damage or described in terms of such damage" (1999, 35).[2] Wall—like Glucklich—takes exception to the necessary linking of tissue damage and pain. Later in his book, Wall suggests that "we should, however, examine the possibility that pain is a syndrome that joins together a coincident group of signs and symptoms, rather than a single phenomenon. It could be that we are aware of the combination of events rather than of pain as a single, separate event" (58). Pain is *marked* as an "unpleasant" awareness: "Pain," Glucklich says, "is conscious by definition" (2001, 96), and another important researcher of pain, William Livingston, has noted that "nothing can properly be called 'pain' unless it can be consciously perceived as such" (cited in Jackson 2002, 148). Toward the end of this chapter, in examining "nonaffective pain," I also note the ways that *remembrance* as well as perception might be a necessary aspect of pain.

It is because of the necessary coincidence of the physiological fact and the phenomenological experience of pain, I suspect, that Elaine Scarry, in her powerful study of *The Body in Pain,* describes the difference between pain and pleasure as the difference between the embodiment of pain and the disembodiment of pleasure (1985, 166). Pain is marked and thereby remembered—even, as we shall see, *materially* remembered—while pleasure is experienced but not remembered with the physiological responses that often accompany pain. On the level of physiology, Damasio argues that pain and pleasure "are different and asymmetric physiological states, which underlie different perceptual qualities destined to help with the solution of very different problems": pain is an after-the-fact "attempt to deal with a problem that has already arisen" (1999, 77, 78)—a problem that has, in fact, *already* marked the body—while pleasure functions "to lead an organism to attitudes and behaviors that are conducive to the maintenance of its homeostasis" (78). This is why Scarry goes to such lengths to describe the sign of the immaterial Hebrew God as himself "marked" by the sign of pain in terms of weapons and injunction: God, like pain, is a monumental matter of fact to be dealt with. Pleasure is also "factual" in this sense—it also marks the body—but, as Damasio says, it "is all about forethought. It is related to the clever anticipation of what can be done *not* to have a problem" (78). In this way pleasure is more obviously related to the semiotics of meaning than pain is. Still, when Peirce defines the symbol as "the law that will govern the future" (1931–35, 1:23) and Hjelmslev defines meaning as "purport" (1961, 55), they are attempting to inscribe within language itself the kind of future orientation that the material markings on our bodies, which usually accompanies pain, create. Meaning, like the stories we tell, builds to some anticipated future, desired or dreaded. But since language and meaning, as

Saussure took such pains to argue, can take up *anything* to articulate its purport—"It is not necessary," he writes, "to have any material sign in order to give expression to an idea; the language may be content simply to contrast something with nothing" (1983, 86)[3]—pain, also, in our experience and anticipation of it, works to create its own future, its own meaning. It might well be that pain is the archetype of memory pushing toward its dreadful or desired termination.

The Physiology and Phenomenology of Pain

The physiology of pain is relatively well understood. It emerges out of the interaction of the three major components of the nervous system: the peripheral nerves, the spinal cord, and the brain. The peripheral nerves include nociceptors, which are receptors that detect actual or potential tissue damage. Patrick Wall notes that

> the skin is profusely innervated with three types of sensory fibres. One group, called A beta fibres, are wrapped in a fatty protein called myelin and are sensitive to gentle pressure. The second group, called A delta fibres, are thinner and are sensitive to heavy pressure and temperature. The third group, called C fibres, are very thin and have no myelin, and respond to pressure, chemicals and temperature. Deep tissue organs such as the heart, bladder and gut are innervated only by the thinner fibres.
>
> . . . Some of our sensory nerve fibres are more than a metre long, running from the toes to the middle of the back; others are only a few centimetres long, running from the teeth to the hind brain. (1999, 40)

The peripheral nerve fibers transmit pain messages to an area of the spinal cord called the dorsal horn in the forms of both electrically induced impulses and—much more slowly (often taking hours or even days [Wall 1999, 41])—chemical transmission. There, chemical neurotransmitters are released that activate other nerve cells in the spinal cord, which almost instantaneously transmit the pain messages up to the brain. More specifically, pain signals are transmitted to the thalamus, a subcortical area of the brain. The thalamus quickly forwards the message simultaneously to three specialized regions of the brain: the somatosensory cortex (the physical sensation region), the limbic system (the emotional feeling region), and the frontal cortex (the region of cognitive activity). Wall notes that "no one area [of the brain] has the monopoly of capturing the one and only input signal associated with pain. One thing is certain: we are not going to find a single pain centre as proposed by Descartes" (56).

Since World War II—and in significant part due to the work of Wall and his colleague, Ronald Melzack—we have learned that the transmission of pain signals is not linear or straightforward. That is, pain messages do not travel directly from the pain receptors to the brain. Rather, when pain messages reach the spinal cord, they encounter specialized nerve cells that act as gatekeepers, which, governed by the brain, filter the pain messages on their way to the brain and even, in the case of weak pain messages, block them entirely. This has been called the gate control theory of pain.[4] In the case of "phantom" pain from missing limbs, the specialized nerves, according to gate control theory, transmit pain messages even when there is no external source of pain (Vertosick 2000, 46–48; see also Wall 1999, 107) in a phenomenon that confuses exteroception and proprioception. In addition, pain messages can change within the peripheral nerves and spinal cord. Nerve cells in the spinal cord may release chemicals that intensify the pain, affecting the strength of the pain signal that reaches the brain. This is called wind-up or sensitization.[5] In these mechanisms, rather than just reacting to pain, the brain actually sends messages that influence the perception of pain. The brain may signal nerve cells to release natural painkillers, such as endorphins or enkephalins, which diminish the pain messages. (This information is based on "How You Feel Pain"; see also Melzack 1993; Damasio 1999, 71–74; Morris 1991, 155–57; Wall 1999, ch. 3; and Melzack and Wall 1983, part 2.)

Subjects without Objects

In the preceding paragraphs, I objectivized and narrativized the phenomenon of pain by making the brain, nerves, and pain messages the subjects or agents of action, just as later I talk about the subjects or agents of feeling in this chapter. Taken as a phenomenon, however, the striking feature of pain is that, as Scarry argues,

> physical pain is exceptional in the whole fabric of psychic, somatic, and perceptual states for being the only one that has no object. Though the capacity to experience physical pain is as primal a fact about the human being as is the capacity to hear, to touch, to desire,

> to fear, to hunger, it differs from these events, and from every other bodily and psychic event, by not having an object in the external world. Hearing and touch are of objects outside the boundaries of the body, as desire is desire of *x*, fear is fear of *y*, hunger is hunger for *z*; but pain is not "of" or "for" anything—it is itself alone. This objectlessness, the complete absence of referential content, almost prevents it from being rendered in language: objectless, it cannot easily be objectified in any form, material or verbal. (1985, 161–62)

Like Saussure in his analysis of language, Scarry refuses to acknowledge the referential "aspect" of pain, to use Wittgenstein's term (2006, 166–82), its location in the world, even while she emphasizes another aspect, the "iconic" sensation of understanding. Thus, some—including myself—might object to this characterization: after all, it is our toe that hurts, and the piano that hurts it, and later I note the anthropologist Valentine Daniel's assertion that, at least at first, pain is overwhelmingly indexical, pointing to the (narrative) situation out of which it arises. Wittgenstein makes a similar point when he focuses on the ways we point to the place that hurts (2006, sec. 298, 302). But Wall notes that the "uncomfortable puzzle" of pain consists in the fact that "stimulus and response seem so strangely connected" (1999, 78). That connection, I argue later, can be understood in terms of material proprioception. The "objectlessness" of pain, its "strange" apparent erasure of relation to its stimulus and, as Scarry says, its seeming "complete absence of referential content," makes the sheer sensation of pain a striking material example of Peirce's icon. The phenomenon of phantom pain reinforces this sense of pain's objectlessness (as does Scarry's focus on torture in examining the body in pain).[6]

The objectlessness Scarry describes makes pain closely akin to Peirce's icon. Moreover, it makes pain more than any other perception or sensation almost purely *embodied:* it marks the body in pain as only itself, without voice, without community or place, without self-consciousness itself. Maureen Flynn notes that "it was this aspect of suffering that Simone Weil recognized when she remarked that pain is the only human sensation capable of completely captivating our attention. 'Here below,' she says about the human condition, 'physical pain, and that alone has the power to chain down our thoughts.' When we experience joy, tiredness, fear, or hunger, she notes that we are capable at the same time of diverting our thoughts in new directions. Under these conditions, it is possible to change our mental channels, so to speak, by an act of will. In intense pain, on the other hand, the human mind can focus on no object other than its own suffering" (1996, 274; for a commentary on Flynn's argument that ties it explicitly to Scarry, see Glucklich 2001, 42–43). Thus, Scarry examines the ways that torture literally effects "the disintegration of the contents of consciousness" (1985, 38), and she particularly argues that "physical pain" has the "ability to destroy language, the power of verbal objectification, [which is] a major source of our self extension" (54). The pain of torture "converts" the realization of civilization represented by any room or shelter "into another weapon, into an agent of pain" (40). "There is nothing contradictory about the fact that the shelter is at once so graphic an image of the body and so emphatic an instance of civilization: only because it is the first can it be the second. It is only when the body is comfortable, when it has ceased to be an obsessive object of perception and concern, that consciousness develops other objects, that for any individual the external world . . . comes into being and begins to grow" (39).

In pain there is no "external world." Scarry makes these observations in the context of her discussion about the objectlessness of God in the Hebrew scriptures, where she argues that God seems only figured in terms of weapons, wounding, injunction. She describes this idea in relation to the "continual reappearance [of pain] in religious experience. The self-flagellation of the religious ascetic," she continues,

> is not (as is often asserted) an act of denying the body, eliminating its claims from attention, but a way of so emphasizing the body that the contents of the world are cancelled and the path is clear for the entry of an unworldly, contentless force. It is in part this world-ridding, path-clearing logic that explains the obsessive presence of pain in the rituals of large, widely shared religions as well as in the imagery of intensely private visions, that partly explains why the crucifixion of Christ is at the center of Christianity, why so many primitive forms of worship climax in pain ceremonies. (1985, 34)

The striking physiological case of objectlessness—as opposed to the "cultural" or semiotic case of religion—is what is called "phantom" pain, sensations, which are often excruciating, that almost all amputees feel in their missing or "invisible" limbs.[7] The pain in phantom limbs has been described with terms such as *jabs, strong current, pins and needles, burning, knifelike, pressure, cramps, crushing,* and *vicelike* (Wall 1999, 11). In fact, the phenomenon of phantom pain led Descartes to argue that pain is felt not by the body but by the soul (see Morris 1991, 157n10; and Glucklich 2001, 53–54).[8] Frank Vertosick also presents a chapter of his book, *Why We Hurt: The Natural History of Pain,* devoted to phantom pain. In his discussion of the physiology of phantom pain, he argues that "the name 'phantom'—[which] is pain in a dead part of the body, the pain of a ghost"—is preferable to "the medical name for this condition . . . , deafferentation pain" (2000, 47). He suggests that "in the absence of any sensation, the spinal cord will feed us static and we will interpret it as burning pain. That's how we've been designed. Pain becomes the 'default mode' of the nervous system. Perhaps," he concludes, "there's some deep metaphysical message in this phenomenon—nothingness equals pain" (48).[9]

Later, I return to the physiological phenomenon of phantom pain in relation to religious suffering. But here let me present a representation of the objectlessness of what I might call "secular" pain. In *The Woman Who Walked into Doors,* a powerful depiction of long-term spousal abuse, Roddy Doyle articulates pain in the first person that approaches the "objectlessness" Scarry describes. Paula is describing—remembering—one of a large number of repeated beatings she suffers from her husband:

> Pushes me, drops me into the corner. Hair rips. A sharper pain. His shoe into my arm, like a cut with a knife. His grunt. He leans on the wall, one hand. His kick hits the fingers holding my arm. I lose them; the agony takes them away. Leans over me. Another grunt, a slash across my chin. My head thrown back. I'm everywhere. Another. Another. I curl away. I close my eyes. My back. Another. My back. My back. My back. My back. Back shatters.
>
> The grunting stops. Breaths. Deep breaths. Wheezing. A moan. I wait. I curl up. My back screams. I don't think, I don't look. I gather the pain. I smooth it.
>
> Noises from far away. Creaks. Lights turned on, off. Water. I'm everywhere. I'm nothing. Someone is breathing. I'm under everything. I won't move; I don't know how to. Someone's in pain. Someone is crying. It isn't me yet. I'm under everything. I'm in black air. Someone is crying. Someone is vomiting. It will be me but not yet. (1996, 183–84)

The pain Paula is subjected to is objectless; it is even subjectless even though it is thoroughly conscious. Like the God of theology, it is nowhere and everywhere. It is, as Scarry says, pure and absolute experience, objectless, iconic, a species of deafferentation pain. Paula's suffering, like the religious suffering of Saint Teresa described later in this chapter, is, as Scarry says, "unlike any other state of consciousness. . . . It is precisely because it takes no object that it, more than any other phenomenon, resists objectification in language" (1985, 5).

Physiological and Narrative Memory

Because pain and our responses to pain involve physical, emotional, and intellectual processing, it is much more complex than the simple stimulus–response model that mechanistic behaviorism created in the early twentieth century. Still, there is an aspect of pain—but not its essential or defining aspect, as the reductionist program of the behaviorists assumed—that renders it simple and meaningless. The memory of pain has been shown to be a *biological* process in cellular mechanisms through which pain is "remembered." Pain itself, as I have suggested, stimulates cellular mechanisms that allow even brief, noxious stimuli to persistently alter the material state of the nervous system, which may lead to central sensitization within the dorsal horn, the brain stem, and the brain. In even primitive organisms, Philip Hilts argues, "external impressions are laid down internally, leading to predispositions, tendencies to feel or react in a certain way" (1995, 28). By "laid down," he means actually, materially *marked* in the body. Eric Kandel—who won the Nobel Prize in Physiology or Medicine in 2000 for his work on the mechanics of memory storage in neurons—has pursued experiments "focusing on a single sensory neuron and a single motor neuron." His research group found that, as in nature, "a shock to the tail [of the giant invertebrate sea snail, aplysia] activates modulatory interneurons that release serotonin, thereby strengthening the [synaptic] connections between sensory neurons and motor neurons" (Kandel 2006, 254). This "strengthening" is, in fact, "the growth of new synaptic connections, an anatomical change that did involve the synthesis of new protein" (256). In other words, Kandel demonstrated the ways in which noxious stimuli led to the physiological alteration of a neuron: neurons are materially "marked" by new growth. This work is part of Kandel's study of the biochemistry of long-term memory. "The ability to grow new synaptic connections as a result of experience," he concludes, "appears to have been conserved throughout evolution. As an example, in people, as in simpler animals, the cortical maps of the body surface are subject to constant modification in response to changing input from sensory pathways" (276). In other words, part of pain is the *anticipation* of pain: the body creates a kind of physiological "memory" of pain. (The natural selection of the physiological phenomenon of material "memory" parallels the natural selection of religious phenomena I touch upon later in this chapter.)

The physiological memory of pain can create all sorts of problems: it creates anticipated pain that disorganizes stable muscular activities, "favoring" a sore ankle as we say, in our physical interactions with the world or disrupting language in our verbal interactions with the world. Talking about the accounts of illness that patients bring to physicians, Dr. Rita Charon and Maura Spiegel argue that

> narratives that emerge from suffering differ from those born elsewhere (unless one argues that all of the business of existing is, to some extent, suffering). Not restricted to the linear, the orderly, the emplotted, or the clean, these narratives that come from the ill contain unruly fragments, silences, bodily processes rendered in code. The language is deputized to point to things not ordinarily admitted into prose or poetry or text of other kinds—shameful, painful, prelingual limitations, absences, breath-taking fears. (2005, vi)

Finally, at a remarkable extreme, the physiological memory of pain can create the dissociated nociceptive neurons that induce the phantom pain of a severed limb (Vertosick 2000, 46). This is important because it creates—as I argue narrative does—a kind of contrary-to-fact phenomenon that confuses the phenomenology and physiology of pain. The very absence of a limb often creates persistent pain that, as David Morris says, "sometimes . . . last for decades. For example, amputees may feel the fingers of the missing hand turned inward and digging into the palm. Absent toes seem cramped ('bunched up'). An entire missing leg can feel icy or burning. A nonexistent wedding ring may still supply its reassuring pressure around a nonexistent finger" (1993, 153).[10] This physiological phenomenon has a parallel in cognition and semiotics. Scarry describes contrary-to-fact discourse—she calls it "counterfactuals"—as the antithesis to the body in pain, even though it often grows out of such pain. "In the long run," she writes, "we will see that the story of *physical pain* becomes as well a story about the expansive nature of human *sentience,* the felt-fact of aliveness that is often sheerly happy, just as the story of *expressing* physical pain eventually opens into the wider frame of *invention.* The elemental 'as if' of the person in pain ('It feels as if . . . ,' 'It is as though . . .') will lead out into the array of counterfactual revisions entailed in making" (1985, 22). As we shall see, this is especially important in relation to religious suffering.

Nevertheless, pain lends itself to be understood as the simple physical facts David Chalmers posits (1996, 41), as "particulars, isolated from their contexts and immune from assumptions (or biases) implied by words like 'theory,' 'hypothesis,' and 'conjecture'" in Mary Poovey's description (1998, 1). That is, the physiological *inscription* of pain in the body itself also—in fact, often—makes the memory of pain, which is what neurologists call episodic memory or explicit memory, function as the opposite of episodic memory: "procedural" or "implicit" memory, the virtually unconscious memory we have for riding a bike, driving, even the process of reading a book. This occurs when we unconsciously favor a once-injured limb with the same lack of intentionality with which we balance ourselves on a bicycle: pain seems simply *there,* a matter of fact. Kandel prefers the terms *explicit* and *implicit* memory rather than *episodic* and *procedural* memory, in large part, I believe, because his more or less mechanistic account of memory rarely takes in the complications of *narrative* memory. Merlin Donald, who as we have seen studies the evolution of consciousness—a very different discipline from Kandel's reductionist science—prefers the term "episodic memory" (see 1991, 151) precisely because he is concerned with the role of narrative in cognitive activity.[11] The physiological memory of pain, of course, is not narrative memory insofar as its physiological alternations are simply—and *meaninglessly*—marked in the body itself. It is meaningless in Peircean terms by not suggesting a law that will govern the future but by simply being a present state of affairs. The marking of the body, in technical terms, does not achieve the double articulation of language that I discussed earlier. Yet almost as quickly as the electrically induced impulses of the peripheral nervous system give rise to neurotransmitters in the spine, so does the biological memory of cells give rise to the episodic memory of suffering people: pain becomes the combination of iconic, indexical, symbolic "modes of being" Peirce describes.

There is good evidence that more than one kind of memory can help us understand and situate the extreme case (and seemingly simple "fact") of sensate pain in relation to the materiality of meaning. In 1962, Brenda Milner began her long-term study of her patient, H. M., who—like the protagonist of the film *Memento*—had lost all long-term memory as a side effect of an operation that removed his hippocampus in order to bring to an end frequent and violent epileptic seizures (see Hilts 1995 for a full-length account of this terrible case; see Sacks for a case history that describes similar memory loss in relation to music [2007, 187–213]). In the course of her thirty-year study of H. M., Milner demonstrated the existence of "more than one kind of memory." "Specifically," Kandel notes in his discussion of this case, "Milner found that in addition to conscious memory, which requires the hippocampus, there is unconscious memory that resides outside the hippocampus and the medial temporal lobe" (2006, 129). Implicit or procedural memory, Kandel argues, is "not a single memory system but a collection of processes" that are akin to the implicit memory of simple animals, including invertebrates, "for habituation, sensitization, and classical conditioning" (132); implicit memory is more or less automatic. Explicit or episodic memory, he suggests, resides (so to speak) in the hippocampus, and it is associated with the multimodality of spatial location. It is striking that Kandel lists pain among the modalities of sensation: "The neurons in the hippocampus of the rat," he notes, "register information not about a single sensory modality—sight, sound, touch, or pain—but about the space surrounding the animal, a modality that depends on information from several senses" (282).[12] Finally—and perhaps most importantly—as the case of H. M. suggests, the hippocampus is the site of long-term memory and consequently a sense of self that is constant through time. (The relationship between a sense of self and an awareness of remembered spacial location—the "highly developed spatial memory system" of primates Emery and Amaral describe [2000, 170]—is fascinating and provocative.) The phenomenology of constancy in time—that one thing follows another and that these events can be grasped together as a meaningful whole—emphasizes the anticipatory nature of memory and its relation to the vastly different *automatic* anticipatory nature of "classical conditioning" that Kandel has studied at length.

In these observations, I argue that pain exists as an episodic or explicit memory—it is something that "happens" to a person—*and* it can also produce the seemingly simple fact of procedural or implicit memory, such as the automatic favoring of a limb, the tightening of back muscles even beyond the healing time of an injured disk, at an extreme the dissociative neurons in the spine or brain or at the point of amputation that induce the phantom pain of a severed limb (for these different "locations" of phantom pain, see Vertosick 2000, 46; Glucklich 2001, 54–57; Wall 1999, 108). In this way the physiological fact of pain seemingly abstracts or isolates the element of *hurt* in the physical processing of memory—the impulse from thalamus to the amygdala that Kandel describes (2006, 345)—from its simultaneous emotional and cognitive processing by means of the hippocampus.[13]

Episodic Memory, Evolution, and Religion

The existence of episodic memory has clear evolutionary benefits. Scott Atran describes the "evolutionary origins of the self" and the degree to which episodic memory contributes to that. The very idea of self or "personhood," as I discuss it later—the very idea of subjectivity as it is articulated in literary texts—is crucial to understanding suffering (as opposed, but also related, to pain), so the degree to which it can be comprehended on the level of biological adaptation will help us grasp the materiality of both pain, religious suffering, and subjectivity as they manifest themselves on the levels of physiology, biology, and semiotics. "In being able to purposefully retrieve episodic memories," Atran writes,

> people are able to (1) access and assess the sources of present knowledge for accuracy and reliability, (2) creatively combine and recombine familiar events to explain the present or anticipate the future, and (3) dissociate or maintain different mental states and scenarios simultaneously. The metarepresentational time-traveling self can thus imagine itself in indefinitely many alternative worlds so as to best choose among them. (2002, 38–39)

Metarepresentational time travel, the maintenance of different mental scenarios simultaneously, and the imagining of alternative worlds are, of course, contrary-to-fact discursive representations.[14] Such phenomena, Pascal Boyer argues, are part and parcel of human life: "Indeed," he writes, "human minds are remarkable in the amount of time they spend thinking about what is not here and now" (2001, 129). He also notes the ways in which such contrary-to-fact thinking is "decoupled" from standard inputs and outputs: one does not see the roof tumbling down when one worries about it (input), and neither does one run to escape (output) (129). Such "decoupled cognition," like toolmaking (chapter 4) and the metarepresentational activity of creating different scenarios simultaneously, is also closely related to religious belief itself, since religion is most often defined in relation to "a supernatural agent or agents" (Dennett 2006, 9), which by definition are *unnatural,* contrary-to-fact, and require the simultaneous entertainment of natural and supernatural phenomena. Thus, Daniel Dennett says, "The core phenomenon of religion, I am proposing, invokes [supernatural] gods who are effective agents in real time, and who play a central role in the way the participants think about what they ought to do" (11–12).[15]

Such supernaturalism helps delineate the two evolutionary "situations" that lie at the heart of most religious phenomena: the situation of finding ourselves in a world where pain and suffering are ubiquitous and the situation of finding ourselves in a world in which the struggle for survival gives rise to "counterintuitive" accountings of experience (whether it is of suffering, danger, or more general cognition). John Bowker describes the first of these situations, noting "the crucial importance in any religion of the account it gives of suffering." "To talk of suffering," he argues, "is to talk not of an academic problem but of the sheer bloody agonies of existence, of which all men are aware and most have direct experience. All religions take account of this; some, indeed, make it the basis of all they have to say" (1970, 2). One cannot say that "the sheer bloody agonies of existence" are to be accounted for by natural selection. Rather, one might say the agonies of existence give rise to natural selection. Still, suffering and a belief in supernatural agencies are closely related, and many have argued that the latter can be accounted for in terms of natural selection.

Atran does so in describing the second evolutionary situation of religious experience. "All religions," he claims, ". . . involve counterintuitive beliefs in supernatural beings," and he contends that "such beliefs are systematically counterintuitive in the same basic ways" (2002, 9). This is part of his larger argument—based on evolutionary cognition—that "in all cultures, supernatural agents are readily conjured up because natural selection has trip-wired cognitive schema for agency detection in the face of uncertainty. Uncertainty is, and likely will always be, ubiquitous. And so, too, the sort of hair-triggering of an agency-detection mechanism that readily lends itself to supernatural interpretation" (71; see also 77). "In sum," he concludes, "it is not an infant-mother, infant-father, or infant-family template per se from which God concepts extend, but a more encompassing *evolutionary program for avoiding and tracking predators and prey.* It is *an innate module for detecting agency and intention, whether good or bad*" (78). Like Atran, Boyer argues that "agent-like concepts of gods and spirits are . . . *natural.* This 'naturalness,' results from the fact that our agency-detection systems are biased toward overdetection," and he further argues that "there are important evolutionary reasons why we (as well as other animals) should have 'hyperactive agent detection,'" since we are creatures "that must deal with both predators and prey" (2001, 145). Atran and Boyer, along with Dennett, David Sloan Wilson, and Richard Dawkins, are anxious to situate religious experience and institutions within the framework of evolutionary cognition. Wilson situates religious phenomena within "cultural evolution," which he describes (following Henry Plotkin) as a "Darwin machine." Such "machines" respond and adapt to changing environmental problems not only by trial and error but also by rational thought conceived of as "a Darwin machine, rapidly generating and selecting symbolic representations inside the head" (Wilson 2003, 31; see also Plotkin 1994).

The ability to entertain different scenarios simultaneously seems to have a material basis in the physical organization of the brain itself; it is the general case of the phenomenon of "theory of mind," discussed in chapter 4 and at length later in this chapter. Studying the function of the right hemisphere of the brain, Robert Ornstein points out that people with damage to the right hemisphere cannot maintain different mental scenarios simultaneously. "In a brain functioning normally," he writes, "while the left hemisphere makes a rapid selection of a single meaning for a word, the right hemisphere keeps in mind a whole range of meanings associated with different situations" (1997, 136).[16] More generally, he observes that "as many researchers in the field have now concluded, the role of the right hemisphere seems to involve maintaining the alternative meanings of ambiguous words in immediate memory, while the role of the left hemisphere is to focus on only one meaning" (110). Ornstein offers the example of a woman who "brought pen and paper with her to meet [a] movie star." An easy first assumption is that the woman is seeking an autograph. When new information is added indicating that this is not her goal (e.g., "her article is going to include famous people's opinions on nuclear power"), most people will revise their expectations and conclude that the woman is a journalist, but patients with right-hemisphere damage "find it very difficult to process this sort of change, and when they retell the sequence, stick with their first, autograph hunter, interpretation" (109). More generally, Ornstein argues that "human beings possess a system whereby one side of the brain performs step-by-step thinking, which we associate with the highest form of human achievements, and the other makes quick judgments of other animals' expressions and perhaps their intentions. The principle of dividing the brain's information processing into two distinct types seems to be both deep and deep-rooted" (28). These two "types" of processing are isomorphic with procedural and episodic memory. Finally, Ornstein notes that the right hemisphere is particularly powerful in religious experiences: "An emphasis on activities of the right hemisphere . . . is the way many of the esoteric Christian, Jewish, Sufi, and other mystical traditions operate. They listen to low tones in chants, view spatial diagrams, puzzle over phrases that have no rational meaning, and attempt exercises to produce a state of 'no conceptualizing while remaining fully awake'" (164; for neurological accounts of these phenomena, see Aquili and Newberg 1999; Newberg and Aquili 2001; and Atran 2002, ch. 7). The "lessening of the verbal and conceptual approach to the world"—which, as Ornstein says, suppresses more or less the functioning of the left hemisphere and encourages "factors that involve the right hemisphere" (165)—might well be linked to the fact, as Scarry argues, that the body in pain is, significantly, without language. In any case, pain also presents the subject with "no rational meaning" and can lead to "a state of 'no conceptualizing while remaining fully awake.'"

In a similar fashion, the ability to comprehend phenomena in terms of "episodic" narrative and narrative structures suggests a material base for subjectivity. Boyer argues that episodic memory is a result of the evolutionary development of humans: "the narrative drive," as he calls it,

> is embedded in our mental representation of whatever happens around us. Also, human are born *planners*, our mental life is replete with considerations of what may happen, what will result if we do this rather than that. Having such decoupled thoughts may well be an adaptive trait, allowing a much better calculation of long-term risks than is available to other species, but it also implies that we represent vastly more life-threatening situations than we actually experience, and that the prospect of death is a very frequent item in our mental life. (2001, 204)

Bruno Latour describes such "decoupling" in his discussion of scientific work in general and that of Albert Einstein in particular. In this, he follows the narrative semiotics of Greimas, who describes the processes of "shifting out" and "shifting in" in relation to narrative comprehension. Latour's and Greimas's terms offer a linguistic framework for understanding Boyer's decoupling: shifting out is the process in language that shifts attention away from the enunciatory situation of discourse—away from the "here and now" occasion for language—toward the more or less abstract "content" of language: it "shifts" away from Peirce's index to his symbol. (For a detailed discussion of Latour's argument, see Schleifer, Davis, and Mergler 1992, 140–47.) Shifting in returns us to that situation that I described in the Introduction as the *worldly* work of literature.

Shifting out allows for what Ornstein describes as the material base for the creation of "a new state of mind" that is derived from the co-evolution of brain and language and, in particular, the specialization of the cerebral hemispheres (1997, 39; in this particular instance, Ornstein is describing the effects of alphabetic writing on human experience). In Greimas's terms, the left hemisphere emphasizes the shifting out or decoupling to achieve abstract "information" while the right hemisphere emphasizes the shifting in toward the here-and-now of changing contexts of enunciation. For Ornstein, the material fact is the advent of writing and literacy that converted intercourse and discourse—the gesturing and oral discourse Donald describes as evolutionary moments—into a "language."[17] Describing the advent of vowels in ancient Greek writing, Ornstein writes that

> the use of the alphabet converted the Greek spoken tongue into an artifact, thereby separating it from the speaker and making it into a language—that is, an object available for inspection, reflection, and continued analysis. It could be rearranged, reordered, and rethought to produce forms of statement and types of discourse not previously available because they were not easily memorable. (39)

Ornstein is describing language in general—and, in his focus on rearrangement, reordering, and even rethinking, the *notational* aspect of language I mentioned in chapter 1—but Greimas's terms apply to narrative discourse and the ability narrative has to create a world that is seemingly separate from that of its speaker, a world that is, to some degree or other, contrary to fact.

Of crucial importance to Latour's (1988) argument about Einstein is that such a narrative is, at least in theory, *writable* (which is to say, subject to notation): that it allows sections of experience or narrative to be detached—decoupled—from the narrative whole and superimposed on other accounts of the world. Shifting out and its complementary activity of shifting in to return to the situation of enunciation—which is, parenthetically, the motor of the psychological sciences, and especially psychoanalysis, as well as speech–act theory—make, as my colleagues and I argue in *Culture and Cognition,* "the 'effect' of cognition possible" (1992, 142), the very sense we have of "knowing" the world, "the social institution of knowledge" (143).[18] They also make toolmaking, outside dependence "on immediate environmental reinforcers or contingencies" (Donald 1991, 179) possible.[19] In terms of religious experience, shifting in allows someone like Saint Teresa of Avila to apprehend her visions of pain and ecstasy—and, indeed, her apprehension of herself as a "person"—as actual occurrences in the world. Teresa is a particularly good case, because she continually questioned her own visions, worrying that they were, in fact, traps set by Satan. Her concern is a remarkable case of acknowledging ambiguities (as Ornstein suggests, a right-brain phenomenon) and a continual process of shifting out and shifting in as well.

The Secret Life of Narrative

Pain and suffering can also be understood in relation to shifting out and shifting in. Almost always, patient narratives, as Charon and Spiegel say, arise in suffering where the situation of enunciation—the *indexical* situation—is markedly pronounced. Patients present themselves because they are uneasy—"dis-eased"—about something: an odd feeling in the chest; disruptions of regular functions such as sleep, sexuality, or speech patterns; unusual fatigue; or unusual loss of muscular control. In this observation, I distinguish between pain and suffering more minutely than Frank Vertosick does (though hardly more minutely than Simone Weil meditates on the religious significance of pain in this chapter's epigraph). More specifically, I follow Eric Cassell's powerful definition of suffering in *The Nature of Suffering and the Goals of Medicine.* The goal of medicine, Cassell argues, is above all the relief or amelioration of suffering: the "test of a system of medicine," he asserts on the first page of his book, is "its adequacy in the face of suffering" (1991, vii). Suffering, Cassell argues, occurs with the destruction or the threat of destruction of those things that define us as persons, the "uprooting of life," as Weil says. Suffering has a past and a future: it is necessarily conscious, it takes place and is apprehended over time, and it exists in the context of the lived life of its subject. Cassell offers a detailed—but perhaps not an exhaustive—list of the characteristics of personhood that are threatened, uprooted, and destroyed in suffering:

> Persons have personality and character, a lived past, a family, a family's lived past, culture and society, roles, associations with others, a political dimension, activities, day-to-day behaviors, and existence below awareness, a body, a secret life, a believed-in future, and a transcendent dimension. (160)[20]

What is remarkable about these characteristics is that most, if not all, of them can be defined or described only through episodic memory that is embodied in and experienced by means of narrative: a lived past, social and political associations, subconscious or unconscious behavior, a sense of the future, even a transcendent dimension—all these categories possess a progressive and graspable existence in time. Suffering, Cassell says "has a temporal element" (36).

What is also remarkable about these characteristics is that, to a greater and lesser extent, they are destroyed by pain. In *The Body in Pain,* Scarry offers a powerful catalog of the "felt-experience of patient or prisoner" of the attributes of pain. Scarry lists eight attributes: (1) the "sheer aversiveness" of pain, the experience of (2) erasing and (3) asserting the agency of the producer of pain, (4) "an almost obscene conflation of private and public" as pain dissolves "the boundary between inside and outside," (5) "its ability to destroy language," (6) "its obliteration of the contents of consciousness," (7) "its totality" ("pain begins by being 'not oneself' and ends by having eliminated all that is 'not itself'"), and (8) "its resistance to objectification" and "representation" (1985, 52–56). These attributes are the direct opposite and negation of Cassell's description of "personhood": its possession of conscious memory of the individual and familial past, its existence within an external world of culture (including associations, roles, politics), its maintenance within ordinary day-to-day activities, and its more or less "unconscious" existence (including bodily life, a contrary-to-fact "secret" or fantasy life, anticipations of worldly and "transcendental" futures). Pain destroys the opposition between outside and inside, the possibilities of conscious life and even *any* rhythms of life beyond its totalizing aversiveness; and it obliterates the human sociality of language and the sense of placement in the world that the objectifications—the shifting out—of language allow, even while it creates an unarticulable sense of overwhelming agency, powers that overwhelm the self and personality itself.

Perhaps Cassell's category of "secret life"—which include "fears and desires, love affairs of the past and present, hopes or fantasies, and ways of solving the problems of everyday life known to only the person" (1991, 42)—is a defining category, because it so clearly involves subjectivity and narrative action that, like "a believed-in future" and a "transcendent dimension," exists as contrary-to-fact discourse, the "law" that will govern the future. Such discourse, I argue, is at the heart of narrative memory: it starkly presents and grasps experience in relation to a meaningful whole that is, like personhood itself, beyond the mechanical operations and facticity of fact. As I have argued, grasping the "meaningful whole set forth by a message" (Greimas 1983, 59) is beyond the mechanical operations, reductiveness, and the simple apprehension of the facticity of fact. Still, the most seemingly mechanical and "factual" phenomenon is the fact of pain itself: this, after all, is the import (the purport?) of Scarry's catalog. And this is why, I think, pain destroys the sense of personhood that Cassell describes, and why it is the locus where the functioning of subjectivity can be most clearly discerned. Personhood, in Cassell's description is a function of narrative comprehension—the comprehension of episodic narrative.

Moreover, even the more detailed neurophysiological descriptions of what Melzack calls "body self" and Glucklich calls "ego" are similarly comprehended in narrative. Relatively late in his career, Melzack went beyond his gate control theory to discuss the material functioning of the felt sense—the phenomenon—of "body self" by which "the body is perceived as a unity and is identified as the 'self,' distinct from other people and the surrounding world. The experience of a unity of such diverse feelings, including the self as the point of orientation in the surrounding environment, is produced by central neural processes and cannot derive from the peripheral nervous system or spinal cord" (1993, 621). Generalizing from this theory, Glucklich presents a neurologically based sense of "ego" that is homologous with Cassell's "person":

> ego can be understood . . . as the end or goal (telos) of the system that constitutes the person. By system I mean more than the individual organism: It includes the family, friends, church, and even inanimate

> objects that surround the individual and define his or her identity in a broad context ("lived world" in phenomenological terms). (2001, 33–34)

More than Cassell, Glucklich emphasizes that aspect of personhood that can be understood as a form of biological adaptiveness: "I am discussing," he writes, "the effects of religiously conceived pain on the development of the self, or the moral agent within his or her community" (34).

Narrative depends on our ability to imagine (contrary to fact?) the intentions of others (or even ourselves) in a series of occurrences that can be grasped as a plotted sequence of events. Dennett calls this the "intentional stance" (2005, 110–11), although most psychologists and philosophers call this "theory of mind" we have already encountered, the imagined or purported "mind" attributed to others. It is this phenomenon that Boyer and Atran as well as Dennett describe as allowing for agency and agency "detection" in their accounts of the ubiquity of notions of supernatural agency across human cultures. It also can help explain the phenomenon of attributing an agent-cause to pain that Scarry describes. More specifically, Atran traces the developmental acquisition of this ability in children in the first years of life, culminating "at the start of the fourth year, [when] the child begins to elaborate an understanding of *metarepresentational agency:* the child attributes intentional attitudes, such as belief and pretense, to people's representations of the world" (2002, 30). The attribution of "mind" to others (and oneself), as I suggested in chapter 4, is the "stuff" of narrative and, indeed, of literature altogether. There I cited Dunbar's contention that "no living species will ever aspire to produce literature as we have it . . . because no other species has a sufficiently well-developed theory of mind to be able to explore the mental worlds of others." Dunbar goes on to describe writing stories as a "fourth-order intensionality" (1996, 102). Pursuing a similar examination of theory of mind, Dennett argues that

> there is some (controversial) evidence that a chimpanzee can *believe* that another agent—a chimpanzee or a human being, say—*knows* that the food is in the box rather than in the basket. This is *second-order* intentionality, involving *beliefs about beliefs* (or *beliefs about desires,* or *desires about beliefs,* etc.), but there is no evidence (yet) that any nonhuman animal can *want* you to *believe* that it *thinks* you are hiding behind the tree on the left, not the right (*third-order* in-

> tentionality). But even preschool children delight in playing games in which one child *wants* another to *pretend* not to *know* what the first child *wants* the other to *believe* (*fifth-order* intentionality): "You be the sheriff, and ask me which way the robbers went!" (2006, 111)

In this passage, the "order" of intentionality is a function of the italicized verbs. Dunbar, in depicting the fourth-order intensionality of literature—but really of complicated narrative in general—describes "writing stories whose plots involve both the writer and the reader understanding [1] what one character thinks [2] another character wants [3] the first character to believe [4]. Since both writer and reader become part of the chain of intensionality, they must be able to go one order beyond what the characters actually do." Dunbar further asserts that "the ability to detach oneself from the immediacy of one's experiences is also a prerequisite for two other unique features of human behaviour, the phenomena we know as religion and science" (1996, 102).[21]

As we have seen, Greimas calls such detachment shifting out; it is the interface of Peirce's index and symbol. Dunbar's inclusion of both writer and reader in the chain of intensionality, moreover, emphasizes the semiotic "disorder" I described in chapter 2 and elsewhere where elements are both within and outside a particular system. The example I present there is the "number" zero, but a more telling example is the "observer" of phenomena that I have repeatedly referred to in discussions of quantum physics. Above all, narrative articulates a series of events grasped as an intentional whole in which part is related to part. Its secret is that it trafficks in agency: shifting in to its events *grasped* as a whole itself "becomes" part of narrative in what Greimas (modifying Vladimir Propp's category of "dispatcher") describes as the relationship between the sender and receiver of discourse, actors or characters who are both within a tale and outside (or really, before) the relations among the actors or characters of a narrative, the shifted-in situation of enunciation or storytelling itself. In *Hamlet,* for instance, the hero's friend, Horatio, "receives" the kingdom, but he also is enjoined to retell the story and functions as the "receiver" of the narrative as a whole, returning the audience to its situation as the recipient of narrative. Thus, in his formal, "structuralist" account of narrative discourse, Greimas includes the material, *worldly* situation of enunciation by means of these narrative categories, sender and receiver, which also characterize different *kinds* of facts, the (shifted-out) story and the (shifted-in) act of storytelling itself, and in so doing he describes *and* negates the hierarchy among the units of discourse.[22] This, then, is the "secret" of narrative: whatever stories they tell, narratives also instantiate and realize the work of agency in the world in their very telling.

Pain and the Materiality of Spirituality

Particularly striking are narratives of pain, because pain and suffering, as Cassell suggests, can be configured to describe and make articulate a "secret life." In one powerful example, Jean Stafford's story of a woman who recently had an automobile accident offers a remarkable verbal-narrative description of pain that is implicit in the visionary experience of Saint Teresa of Avila, a sixteenth-century Spanish mystic from whom Stafford takes the title of her story "The Interior Castle." In fact, Stafford's story achieves its articulate sense of pain by suggesting a parallel between the story of Pansy's painful operation after the accident and the writings of Teresa. The title is an explicit connection between the experience of the accident victim and that of the late medieval mystic: Teresa's *The Interior Castle* begins her meditation by noting that "I began to think of the soul as if it were a castle made of a single diamond or of very clear crystal, in which there are many rooms, just as in Heaven there are many mansions," and she describes it as a place in which "the most secret things pass between God and the soul" (Silverio de Santa Teresa 1946, 2:201, 202); and Stafford's story repeatedly contrasts theological vocabularies and scientific vocabularies.

Later I return to Stafford's story, but here I want to examine Teresa's "secret" religious life in relation to pain, suffering, and narrative in order to describe the ways that the physiological *fact* of pain can be can be *phenomenally* gathered up into a meaningful whole of religious significance and of a concomitant powerful sense of subjectivity. In *The Interior Castle* and elsewhere throughout her writings,[23] Teresa combines the experience of pain and religious vision: Helmet Hatzfeld notes that "all the pains that she was accustomed to in her life, the paralysis and shrinking of her nerves which the doctors used to call unbearable, were nothing compared to those visionary pains which, as she understood, would never subside and which caused her a really intolerable oppression, suffocation, affliction, and despair" (1969, 84–85); and Atran observes, quoting *The Interior*

Castle—in the context of Teresa's possible suffering from epileptic seizures (paralleling, in Atran's account, a similar possibility for the Apostle Paul; see also Trimble 2007, chapter 7, for the "neuro-theology" of epilepsy)—"she experienced vivid visions, intense headaches, and fainting spells, followed by 'such peace, calm, and good fruits in the soul, and . . . a perception of the greatness of God'" (2002, 189). More extensively in *The Interior Castle,* Teresa herself notes that "as I write this, the noises in my head are so loud that I am beginning to wonder what is going on in it. . . . My head sounds just as if it were full of brimming rivers, and then as if all the water in those rivers came suddenly rushing downward; and a host of little birds seem to be whistling, not in the ears, but in the upper part of the head, where the higher part of the soul is said to be" (Silverio de Santa Teresa 1946, 2:234).

In her *Life,* published in 1667, a decade before *The Interior Castle* was published, Teresa offers the most famous of her descriptions of intense visionary religious experience, one that occasioned Bernini's famous sculpture *The Ecstasy of Saint Teresa* and Richard Crashaw's poem "A Hymn to the Name and Honor of the Admirable Saint Teresa," two generations after she died in 1582. In her vision, she is visited by an angel in whose hands she saw

> a long golden spear and at the end of the iron tip I seemed to see a point of fire. With this he seemed to pierce my heart several times so that it penetrated to my entrails. When he drew it out, I thought he was drawing them out with it and he left me completely afire with a great love for God. The pain was so sharp that it made me utter several moans; and so excessive was the sweetness caused me by this intense pain that one can never wish to lose it, nor will one's soul be content with anything less than God. It is not bodily pain, but spiritual, though the body has a share in it—indeed, a great share. (1:192–93)

Afterward—"during the days that this continued," Teresa writes—she says that she "had no wish to see or speak with anyone, but only to hug my pain, which caused me greater bliss than any that can come from the whole of creation" (1:193). Morris emphasizes the "blended, eroticized, mystical pain" both in Teresa's narrative and Bernini's sculpture (1993, 132), but more to the point, I think, is the way in which Teresa's experience instantiates Scarry's sense of the totalizing power of pain: "pain begins," Scarry writes, "by being 'not oneself' and ends by having eliminated all that is 'not itself'" (1985, 54). For Teresa the pain is "sharp" but becomes quickly (immediately?) something one cannot imagine being without. Elsewhere, Teresa repeatedly describes her religious experience as a kind of out-of-body experience, very much in accordance with the ways that Scarry describes the "objectlessness" and ego-destroying effects of pain.[24]

Teresa's pain and sweetness comprehend the full recognition Maureen Flynn describes of "the utter abjection into which the mystics threw themselves in their quest for divine reunion" (1996, 270). "Throughout the rituals of the mystics," Flynn continues, "there can be found the presence of pain—sharp, constant, and heavy pain. In the attempt to eliminate all human sensation an exception was made for affliction—both physical and emotional affliction" (271). In her account of the spiritual uses of pain, Flynn specifically describes Teresa:

> Pain was the one human quality that she carried with her. To be fully embraced by her Bridegroom she had to destroy first the animal passions and physical sensations ruling the organic body. Pain constituted proof that her human nature was being purified. She affirmed suffering along the passage to psychic purity because it suggested that she was nearing her destination to an immaterial presence. In this theology of ecstasy every tinge of corporeality implied the beginning of the end of temporal existence. Physical sensation of the most unbearable sort indicated that the hand of God was tearing from the soul its worldly inclinations. (273)

In an odd way, pain—the *most* corporeal sensation, precisely because with it, as Scarry suggests, there is nothing but body—pain becomes ghostly, immaterial, otherworldly. Here is where the remarkable phenomenon of phantom pain can help our understanding. In this understanding, as Glucklich describes it, "pain, in short, unmakes their profane world and leads the mystics to self- and world-transcendence" (2001, 42).

The phenomenon of phantom pain is both world-transcendent and self-transcendent. "Phantom-limb pain," Glucklich argues, "can be regarded as a type of hallucinatory experience. The pain is absolutely real, of course, but the certainty of its localization in the missing limb is an 'illusion' that results from the absence of input from that limb" (57). Glucklich compares phantom-limb pain to visual hallucinations—which are, he says, "similarly, common among ascetics and mystics who retreat to environments where sensory input is reduced to minimal levels" (57; Sacks describes similar aural hallucinations as "so-called 'release' hallucinations," 2007, 52)—and he refers to the remarkable work of V. S. Ramachandran, who has relieved phantom pain by using mirrors to allow amputees to "see" their missing limbs and thereby unclench them. Placing a mirror between a patient's whole and amputated hands, Ramachandran and his colleagues then asked their patient

> to make symmetrical movements of both hands, such as clapping or conducting an orchestra, while looking in the mirror. Imagine [the patient's] amazement and ours when suddenly he not only saw the phantom move but felt it move as well. I have repeated this experiment with several patients, and it seems that the visual feedback animates the phantom so that it begins to move as never before, often for the first time in years [during which the missing limb continually felt clenched and cramped]. Many patients have found that this sudden sense of voluntary control and movement in the phantom produces relief from the spasm or awkward posture that was causing much of the agonizing pain in the phantom. (2004, 16–17)

Glucklich claims that Ramachandran's work demonstrates that pain is significantly a function of cerebral rather than peripheral nerves and that the model of a cerebral "neuromatrix" developed by Melzack can account for Teresa's sense that pain is both excruciating and "sweet." (For a fuller account of the phenomenology of sacred pain, see Glucklich 2001, especially chapter 4.)

Here is how Melzack's model works. The phenomenon of what Glucklich calls sacred pain is closely related to Cassell's sense of a secret life. In fact, Glucklich describes a sense of "personhood" in terms of phantom pain and the "body-self template" Melzack has posited from the experience of phantom pain (see Melzack 1993, 621–27).[25] Specifically, he describes Melzack's notion of the neuromatrix—"the anatomical network that consists of neural loops between the thalamus and the cortex as well as the cortex and limbic system," a "(flexible) systemic structure [that] processes incoming nerve impulses according to a characteristic pattern" (Glucklich 2001, 54–55) in Melzack's theory—which is, he argues (as does Melzack) "the neuropsychological foundation of the phenomenal self" (2001, 58). Marni Jackson also glosses Melzack's theory: "The concept of a neuromatrix," she writes, "suggests that pain is not an invasive, alien force or a learned response but part of the map of who we are. Pain should not surprise us (a point of view that Buddhists have been teaching for some time)" (2002, 76). Basing his surmise in large part on the phenomenon of phantom pain, Melzack postulates a body-self template that is, Glucklich argues, "at the basic information level, . . . a spatiotemporal organization of inputs and outputs relating to the bases of unified perception and proprioception. It provides the structure of the phenomenal self a schema or schemata" (2001, 58). Proprioception is the perception of one's own body—muscle tension (including the weight of objects in our hands), balance, and, in Scarry's description of objectlessness, pain as well. The self, Melzack and Glucklich argue, is the phenomenal mapping of the world and the body.[26] Phantom pain is a function of imbalances of input and output of these systems: "Brain commands," Melzack writes, "may produce the experience of movement of phantom limbs even though there are no limbs to move and no proprioceptive feedback" (1993, 625); and "phantom pain and hallucinations," Glucklich says, "are produced by the absence of incoming neuronal signals. The neurosignature gradually overfires due to the absence of modulating or damping feedback from the extremities" (2001, 57; Glucklich notes that "the neuromatrix processes incoming nerve impulses according to a characteristic pattern, which Melzack calls 'neurosignature'" [55]). Moreover, among other things, the sacred pain Glucklich describes has the effect of *realizing* a secret life, and it does so—in the way semiotics itself does—by means of *absence* creating the significance of positive fact. In the case of Teresa, her "secret life" was the continual but invisible illusory presence of Jesus in her day-to-day life, a "vision" said to have lasted more than two years—beginning, significantly, when she encountered an image of the suffering Christ (Silverio de Santa Teresa 1946, 1:54). This experience, along with the vision of the angel, led her to a cry that was often inscribed as a motto upon her images: "Lord, let me suffer or let me die." Material pain *with all its destructive and negative force* becomes, as Jackson says, "part of the map of who we are" (2002, 76).

In this way, the work of sacred pain, Glucklich argues, is to *realize* personhood. It does so by creating a sense of empowerment (2001, 98) but also, and most strikingly, by diminishing "the subjective experiencer" to the point of creating "a 'reverse phantom'" (59) of a subject absorbed into a higher power, Teresa's "Bridegroom." (This is an example of shifting out.) Glucklich cites Valentine Daniel's remarkable description, based on Peirce's categories, of the Ayyappan pilgrimage in India where pilgrims suffer from lacerated feet, blisters, blistering sunshine in a barefoot forty-mile pilgrimage. "Sooner or later," Daniel (who himself underwent the pilgrimage) writes,

> all the different kinds of pain begin to merge. Your knowledge via Thirdness begins to cease to function. It is replaced by Secondness. The experience of *pain* makes one acutely aware of oneself (ego) as the victim, and the outside (undifferentiated as roots, stones, and hot sand) as the pain-causing agent. This is again Secondness. . . .
>
> Pain also has an element of Firstness. . . . With time, pain stops having a causative agent, and ego is obscured or snuffed out because it has nothing to contrast itself with or stand against. Ego is no longer a victim, as in Secondness, because the identity of the pain-causing agent is lost. There is a "feeling" of pain, of course, but it is a sensation that has no agent, no tense, and no comparative. One does not know whence the pain came, how it is caused, or whether it is more or less intense than the pain a moment or an hour ago; for there is no before or after. Pain is the only sensation belonging to the eternal present. (1987, 267–68)

Here, as in Teresa's pain, the objectlessness of pain overwhelms personhood in order (paradoxically) to delineate a secret life. Daniel adopts Peirce's categories of Firstness, Secondness, and Thirdness—categories that grow out of the functioning of the sign, which is First a *representamen* ("the artifact, sign vehicle, or event" [Daniel 1987, 16], that is, a sensuously apprehended phenomenon), Second an *object* (an "empirical" fact, though it "need not be a material thing" [18; for the identification of object and fact, see 19, 30]), and Third an *interpretant* ("the locus of interpretation, that by which a sign is contextualized, that which makes signification part of a connected web and not an isolated entity" [18]). Moreover, these three categories encompass the Peircean categories of signs—what Daniel describes as the "relationship of sign to object" (30)—discussed in the previous chapters: the icon, index, and symbol. Index, as we have seen, indicates empirical facts.[27]

In part, Daniel uses Peircean categories to describe the relationship between pain and religion (though his larger aim is to develop a systematic working definition of "culture"). Thus he claims that "nerve impulses and other perceptual judgments are almost exclusively indexical signs, that is, signs whose significance is a function of the necessary contiguity of the *object* and representamen" (26). He also suggests, quoting Clifford Geertz, that a significant part of religious beliefs is its indexical function. "*Religion,*" Geertz writes, "*is (1) a system of symbols which acts to (2) establish powerful, pervasive, and long-lasting moods and motivations in men by (3) formulating conceptions of a general order of existence and (4) clothing these conceptions with such an aura of factuality that (5) the moods and motivations seem uniquely realistic*" (Geertz 1973, 90). Daniel's larger example, however, is the Tamil Ayyappan pilgrimage during which the overwhelming *indexicality* (Secondness) of pain "replaces" the *symbolic* Thirdness of knowledge and thought and is itself replaced by the *pure* (iconic) experientiality of objectless pain (Firstness). More specifically, he claims that the Hindu pilgrim strives "to move away from the world of theories and laws, through Secondness, which dissolves the rules and theories that classify 'the other' into just one other (in the case of the pilgrimage, *pain*), which looms out against the self, to Firstness, where the 'other' is experienced in such immediacy that it is no longer clear where the self ends and the other begins" (1987, 244).[28] In this experience, as in Teresa's vision, "pain itself [is] conquered by love" (257). "Several pilgrims seemed to believe, however," Daniel concludes, "that after a while, pain, having become so intense, began to disappear. In the words of one pilgrim from my village, 'At one moment everything is pain. But the next moment everything is love *(anpu)*. Everything is love for the Lord'" (269).

There is a more specific—perhaps a more "material"—way of describing the objectlessness of pain that Daniel describes. The nociceptive activity of pain receptors—the physiology of pain—participates in both the "exteroception" of external phenomena and the "interoception" of distress of deep body organs (such as the heart or the stomach). That is, Glucklich argues, there are two basic functions of pain perception. "The first function," he writes,

> which is localized in the A beta fibers, "is a warning system that provides immediate information about the presence of injury, the extent of injury and its location." The second function, which is localized in the C fibers, is a reminding system: "By generating slow, diffuse, particularly unpleasant and persistent pain, the second system repeatedly reminds the brain that injury has occurred and hence that normal activity should be restricted." The first system is based on extero-

> ception and the second on interoception. The sharp bright pain of the first function generates an immediate muscular escape response, the hand that quickly withdraws from the hot stove, for instance. The dull ache of the second function produces inactivity and avoidance of unnecessary motion. These responses are natural and characteristic of animals as well as humans. (2001, 71; Glucklich cites Chapman 1986, 158; see also Wall 1999, 98–99)

When Scarry describes objectless pain, she is describing the *phenomenon*—the felt experience—of proprioception superseding perception in nociceptive activity so that self-referential bodily action, such as pain or simply touching, seems to be a state of the body unrelated to any particular object in the world (whether it be a hot stove or the ankle's persistent soreness). In a similar fashion, Daniel notes that "pain, to begin with, is multiple and differentiated"—the pain of toe and heel blisters, of the weight of his backpack, of headache in the baking sun (see 1987, 272)—but "with time, the various kinds of pain merge into a unitary sensation of pain" that is then replaced by "love" (274). As such, it seems that the phenomenally overwhelming experience of the integrity of what now seems a transcendental self, governed by a body-self template, feels as if it is absorbed into an impersonal sense of pure being, *iconic proprioception.*

It does so in a remarkable transformation of the significance of pain. "People in pain," Cassell writes,

> frequently report suffering from pain when they feel out of control, when the pain is overwhelming, when the source of the pain is unknown, when the meaning of the pain is dire, or when the pain is apparently without end.
>
> In these situations, people perceive pain as a threat to their continued existence—not merely to their lives but their integrity as persons. That this is the relation of pain to suffering is strongly suggested by the fact that suffering can often be relieved *in the presence of continued pain* by making the source of the pain known [and] changing its meaning. (1999, 36)

The apprehension—the very seeking—of sacred pain works to establish the "integrity" of a subject as a person by establishing her as both the subject and—seemingly impossible for Scarry—the *object* of pain. In the case of Teresa, it transformed the nature of her understanding of her integrity as a person and, as in the negative case of phantom pain, it transformed her very body-self template. She understood, as George Eliot notes in *The Mill on the Floss,* that Thomas à Kempis understood "the inmost truth . . . that renunciation remains sorrow, but a sorrow borne willingly" (1961, 255).

Another way to say this is to return to Scarry's distinction between pleasure and pain. "If a thorn cuts through the skin of [a] woman's finger," she writes, "she feels not the thorn but her body hurting her. If instead she experiences across the skin of her fingers not the awareness of the feel of those fingers but the feel of the fine weave of another woman's work, . . . she . . . experiences the sensation of 'touch' not as bodily sensations but as self-displacing, self-transforming objectification" (1985, 166). Scarry argues that this feeling is called "pleasure," "a word," she says, "usually reserved either for moments of overt disembodiment or, as here, moments when acute bodily sensations are experienced as something other than one's own body" (166). In a note to this passage, she describes pleasure as "a condition associated with living beyond the physical body, or experiencing bodily sensations in terms of objectified content" (355). For Teresa—and for the experience of sacred pain more generally—the pain that is felt, such as when Teresa glimpsed the image of the suffering Jesus or when she feels impaled by an angel, also provokes the pleasure of self-displacement and self-transformation: pain—even the pain of disembowelment—is objectified as a feeling and element integrated into a whole, now transcendental self, "completely afire with the great love for God" (Silverio de Santa Teresa 1946, 1:193). In this way, Teresa, like the Hindu pilgrim, combines ecstasy and pain.

Morris notes that Teresa is a "visionary figure in whom rapture and pain seem strangely united" (1991, 131). In her *Life,* as we have seen, she describes her vision of an angel plunging a dart "deep within me." That "deepest part" is Teresa's "secret self" that, in her visionary expression, combines Cassell's attributes of personhood and Scarry's attributes of pain in realizing sacred pain. This combination, I think, can help us to see the ways that the physiological fact of pain can be taken up by—rather than destroying—the phenomenological fact of personhood, the ways that iconic Firstness is apprehended as symbolic Thirdness and an experience of "pure being" is achieved. In her visionary language, Teresa presents what I call the double articulation of pain. Like the double articulation of language André Martinet described in the 1930s, her narrative presents the signification of suffering in relation to the physiological fact of pain and, in its abstract language (shifting out), the transformation of subject from the "I" who suffers the "sharp pain" to the subject who "can never wish to lose" the sweetness "caused me by this intense pain." I might add, in the terms of Peirce that Daniel adopts, the double articulation of pain that Teresa experiences (enacts?) is the *fact* (or Secondness) of pain and its sensuous, undifferentiated possibility (Firstness).[29] The double articulation of suffering and pain (and I use this term of art from structural linguistics quite literally) relates the almost Cartesian mechanism of the physiology of pain to a nonmechanical semantics. It accounts for the ways that the *material* absoluteness of objectless pain can create the effect of a secret life for its subject that is ghostly, intangible, and otherworldly.

Double Articulation

In his study of Hindu pilgrims, Daniel notes that "in Hindu India, iconicity is valued over symbolization, whereas in the modern West, the quest for indexical and symbolic signs is valorized" (1987, 40). At the end of his study, he repeats this contention: "In this final merging, there are no analytic distinctions to separate self from other (Thirdness); no rude consciousness of the other (Secondness). In Firstness the pilgrim is able to know (if not say), 'Now I really know.' For Peirce, as for the dominant Western tradition he represents, the epistemological movement is from Firstness to Thirdness, from synthetic to analytic knowledge. For the Hindu pilgrim, however, the knowing process moves from Thirdness to Firstness, from analysis to synthesis and then further, to where there is nothing left to know either through analysis or synthesis. This is *pūrna* vidyā, perfect knowledge" (286). This is also true, I suspect, for Teresa. However, in her modern-day story, "The Interior Castle," Stafford counterpoints Teresa's traditional language of "soul" and "redemption" with the scientific–materialist language of "brain" and biomedical science (though Teresa herself ventures into "scientific" description when she talks about the "upper part of the head" as the locus for the higher part of the soul in *The Interior Castle*). Awaiting her operation in the hospital after her terrible automobile accident, "what Pansy thought of all the time," Stafford writes,

> was her own brain. Not only the brain as the seat of consciousness, but the physical organ itself which she envisaged, romantically, now

> as a jewel, now as a flower, now as a light in a glass, now as an envelope of rosy vellum containing other envelopes, one within the other, diminishing infinitely. It was always pink and always fragile, always deeply interior and invaluable. She believed that she had reached the innermost chamber of knowledge and that perhaps her knowledge was the same as the saint's achievement of pure love. It was only convention, she thought, that made one say "sacred heart" and not "sacred brain." (1969, 183)

Stafford implicitly articulates this language in relation to Teresa, who in *The Interior Castle* speaks of the "infused light or of a sun covered with some material of the transparency of a diamond" (Silverio de Santa Teresa 1946, 2:315). But there is a significant difference between the modalities of Teresa's religious experiences of pain and Stafford's representations of Pansy's pain (and her own: Stafford was also a victim of a terrible automobile accident). While Daniel explains that difference in Peircean terms, it can also be understood in relation to the concept of double articulation examined in chapter 4, the "duality of patterning" that Michael Corballis describes as both "a hallmark of true language" and "one of the most difficult problems of language evolution" (2002, 114). Double articulation is also touched upon in chapter 3, where I cite Greimas describing poetry as "the shortening of the distance between the signifier and the signified." Greimas says that "poetic language, while remaining part of language, seeks to reachieve the 'primal cry,' and thus is situated midway between simple articulation and a linguistic double articulation. It results in a 'meaning-effect' . . . which is that of 'rediscovered truth' which is original and originary. . . . It is [an] illusory signification of a 'deep meaning,' hidden and inherent in the [phonological] plane of expression" (1970, 279; my translation).

As we have seen, the double articulation of language describes the systematic development of sign systems: symbols (or semantic content) are systematically organized in relation to material signifiers, sound-images, as Saussure called them, or phonemes. Saussure's sound-image combines (or confuses) the immediate sensuousness of Peircean Firstness and the facticity of Secondness. (This is why, I think, it is proper to talk of the feeling of immediacy in the apprehension of meaning as well as sensual apprehensions.) But, then, pain does this as well: it is overwhelming feeling that, like a fact, cannot be denied. The strength of Saussurean linguistics—but also its weakness, insofar as it seems to banish reference from its understanding of language[30]—is its folding of the Peircean interpretant into the signifier. This gesture, like the painful pilgrimage Daniel describes, accomplishes the "illusion" of the transformation of material fact to seeming intangible phenomenon; it accounts for the *immediacy* of meaning we experience in our understanding. (Even when we figure something out, the final meaning *feels* immediate.) It is for this reason, I think, that for Greimas, poetry creates the illusion of the discovery of "deep meaning," the primal cry of experience itself charged with significance. Yet such an experience—even the experience of pain—charged with significance might well define the sacred altogether or, in perhaps more secular terms, the literary.[31] Daniel describes this in Peircean terms when he describes the way, as we have seen, the Hindu pilgrim moves away from Thirdness, "the world of theories and laws, through Secondness, which dissolves the rules and theories that classify 'the other' into just one other (in the case of the pilgrimage, *pain*) . . . to Firstness, where the 'other' is experienced in such immediacy that it is no longer clear where the self ends and the other begins" (1987, 244). If we were to map Peirce's terms onto Martinet's (and Saussure's), we would note that Thirdness—discourse conceived as symbol—inhabits the first articulation of meaning, while Firstness encompasses the sensuous quality of unmediated experience is combined with a sense of its facticity to achieve the second articulation of language. Lessening the distance between the two, a primal cry of meaning creates the illusion that meaning *inhabits* sensual facticity. Such knowledge (or is it experience?) is what Teresa seeks and, by her own account, achieves. It is not, however, what Stafford's story or her character, Pansy, achieves. Her story achieves the articulation—the *double articulation*—of the experience of pain, a self-conscious sense of its illusory significance. Like the poetry that Greimas describes, Stafford's narrative aims at transforming the rude otherness of pain into its comprehension within our intellectual tradition; and like the indexical poetry of Keats, we can find in her discourse the transformation of narrative into assertions of the immediate (illusory?) phenomenon of subjectivity.

The Material Self

The crucial issue here—it is the crucial issue for Chalmers—is the materiality (it would be the "intangible" materiality) of phenomena, and especially that extreme case of phenomena, the immediate and abiding sense of subjectivity in relation to pain, which alternatively destroys, transforms, and reinforces that sense.[32] I have argued throughout this book that it is possible and necessary to describe intangible materialism in the light of semiotics and information theory and particularly in the light of the very existence of literature as the elaboration of both theory of mind and the contrary-to-fact sense of sacred agency in the world. Moreover, it is equally important to describe phenomenal materiality, as I do here, in terms of its relation—in its material *homological* relation—to the physical and biological sciences. That is, the intellectual comprehensions of pain as a physical "fact" and as a biological response to the world are *related to*—and, in fact, create basic "nestings" for—the material phenomenal experience of self. For this reason, I think, Oliver Sacks, reporting on his patients who had "awakened" after a half century suffering from catatonia, designates the phenomenon that Pansy—and, to a degree, Teresa and Cassell—describe so romantically and metaphorically in a more scientific language. There is, he speculates, as we have seen, an "engram" of personhood, a set of *habitual* responses to experience that, inscribed in the cellular and networked organization of the brain (most likely in the manner that both Kandel and Melzack describe), literally *characterizes* a person. (It is no accident that "habit" is a crucial term in Peircean semiotics: thus Daniel notes that for Peirce "'Man is a bundle of habits' [6.228]. Belief is habits become conscious [4.53]. Becoming conscious of habits is a distinctively human attribute, making reflection, and even critical self-reflection, possible" [Daniel 1987, 25].)[33] "Many neuropsychologists," Sacks writes,

> have spent their lives "in search of the Engram" [that more or less permanence-effect produced by stimulation by means of which] . . . individual skills and memories may survive massive and varied extirpations of the brain. Such experimental observations . . . indicate that one's *persona* is in no way "localisable" in the classical sense, that it cannot be equated with any given "centre," "system," "nexus," etc. but only with the intricate totality of the whole organism, in its ever-changing, continuously modulated, afferent–efferent relation with the world. (1999, 239)

Such characteristics—and characteristic *narrative* relationships with oneself, one's environment, one's contrary-to-fact imaginings—comprise the kernel, the core of personhood, as Cassell describes it. Throughout his work, Sacks studies the *physical,* neurological destruction of these characteristics. And Glucklich asserts that "neuropsychologists, Gestalt theorists, and philosophers of mind argue that the self is an emergent property of several organic systems and is not in fact separate from the body" (2001, 94).

What emerges from organic systems is phenomenological sense: body-image, body-schemata, self itself. Even though "the body-image is a phantom," Glucklich writes, "we do in fact experience ourselves *as though* possessing an integrated and concrete body-self" (90). We have encountered several phantoms throughout this book: phantom pain and phantom primal cries, phantom supernatural agents, phantom self-evident accounting, phantom subjects, across levels of physiology, biology, and semiotics. Glucklich notes the phenomenal "consciousness of a unitary self . . . [which] owes itself to a combination of cultural and biological processes" (91), and he argues—in a manner parallel or isomorphic to my arguments in chapters 2 and 4—that

> the "self" is grounded in a complex hierarchy of organizations that can be described from different perspectives as either organic or psychic. From the biological point of view, the human organism consists of numerous systems and subsystems of cellular organization. At the most basic level are single cells, then individual organs, and moving up the scale there are complex functional systems such as the circulatory system, the reproductive system, the immune system, the nervous system, and so forth. In calling a unit a "system" we are making several assumptions. For instance, the separation into systems is functional rather than anatomical. The same organs can be "shared" by more than one system. The organizational principle implies at least three critical features: communication within the system, the notion of telos or systemic goal, and isomorphism between different levels of subsystems. (91)

Glucklich defines isomorphism, following Gestalt theory—and in a manner that is close to the definitions of homology and analogy with which I began—as "similarities" that can be seen but that can also be understood as "similar" responses to the physical-material environment so that the patterns of strolling of African elephants and human strollers in New York—the "rhythms of leg movements," "the pace and bounce" of walking, computations of "limb size and joint plasticity"—reveal "subtle but discernible biomechanical laws" (93). This, Glucklich argues, leads to

> the fact that intelligent behavior does not require a Cartesian mind [and] that computation and other mental processes are subject in

> principle to a scientific analysis that discards the "little man in the head" theory. The isomorphism that Gestalt psychology identifies between levels of organization is based on the fact that subsystems obey fundamental principles of physics. Antimetaphysical theories may seem overly behavioristic or materialistic to some readers, but they have been very successful at describing and predicting patterns in perceptions and even behavior. . . . I believe that material and cybernetic theory is appropriate for understanding voluntary pain. (94)

In Stafford's story, Pansy does not encounter voluntary pain—in fact, it is precisely her everyday "secret life" that is threatened by what Pansy calls her "two adversaries: pain and [her physician] Dr. Nicholas" (1969, 183)—but nevertheless the narrative attempts to capture the experience of pain's objectlessness. In the story, Dr. Nicholas embodies conventional biomedicine and, indeed, as Stafford's catalog of his attributes suggests—"an aristocrat, a husband, a father, a clubman, a Christian, a kind counselor, and a trustee of his preparatory school" (186)—he also embodies the conventional middle-class life of what Teresa describes as that of a "tepid Christian" (see Hatzfeld 1969, 85). Still, in the story, it is pain that is Pansy's chief adversary, threatening her sense of personhood, and even the possibility of being "absorbed" into a higher power is threatened insofar as Pansy imagines her physical brain the locus of that power. As Stafford notes, during the operation, while Dr. Nicholas proceeds, Pansy is

> in such pain as passed all language and even the farthest fetched analogies. . . . She was claimed entirely by this present, meaningless pain and suddenly and sharply she forgot what she had meant to do. She was aware of nothing but her ascent to the summit of something; what it was she did not know, whether it was a tower or a peak or Jacob's ladder. Now she was an abstract word, now she was a theorem of geometry, now she was a kite flying, a top spinning, a prism flashing, a kaleidoscope turning. (1969, 189)

Pansy's "present, meaningless pain" destroys both the future and the past; it is as timeless as an abstract word, a theorem of geometry, a mechanical operation. It overwhelms her "interior castle," her secret life, and the anticipation of its timeless destruction of her personhood transforms pain into—or rather *realizes* pain as—great suffering, affliction itself.[34]

The parallel of Stafford's description of pain to the narrative of Saint Teresa is remarkable, in large part because of the ways that the physiological memory and anticipation of pain Pansy experiences discovers a vocabulary for its interpretive narration, the repeated unruly fragments, silences, bodily processes rendered in coded figures of speech. Teresa combines ecstasy and pain; Morris notes that she is a "visionary figure in whom rapture and pain seem strangely united" (1993, 131). This combination again demonstrates that the physiological fact of pain can be taken up by—as an alternative to destroying—the phenomenological experience of personhood we saw in Teresa. Stafford's narration presents pain and suffering side by side, even though Dr. Nicholas can see only the positive fact of pain, without memory or dread, simply by itself. Pansy has a secret life—of mind and soul and inscrutable intelligence—which her doctor makes no attempt to comprehend. And she knows—cognitively and emotionally—that anticipating the destruction of conscious memory ("she forgot what she meant to do") is at the heart of her suffering. Yet more explicitly than Teresa—perhaps because she cannot fall so easily into a theologically organized "transcendent dimension" or perhaps because Stafford, unlike Teresa but like Peirce, equates knowledge with Thirdness—Stafford presents the double articulation of pain: pain as meaning but also as material sensuous fact. At the very end of the story—after the operation, during which Pansy recounts narrative after narrative of her family, her mother's friends, and contrary-to-fact summertime birds (the operation takes place in the dead of winter)—back in the hospital room, "there was great pain," we are told, "but since it could not serve her, she rejected it and she lay as if in a hammock in a pause of bitterness. She closed her eyes, shutting herself up within her treasureless head" (1969, 193). Pansy's self is neither the impersonal "I" of the lyric nor the social "I" of narrative nor even the transcendental "I" of Theresa. Forged in and against pain—against "the sheer bloody agonies of existence, of which all men are aware and most have direct experience" that John Bowker describes (1970, 2) and Emmanuel Levinas suggests—Pansy's subjectivity catches up together physiological fact, social narrative, and impossible (contrary-to-fact) semiotic figures, "a pause of bitterness." Pansy's pain only aspires to the status of "sacred," and it is precisely the *materialism* of Stafford's account of pain that allows her to play with Teresa's "interior castle" without fully embracing it. Such materialism is implicit in semiotics as well, in the very fact of its double articulation encompassing the systematic semiosis of meaning and the sensuous fact of embodied discourse.

Memory and Concern

A final way to examine the relationship between pain and suffering—and also to delineate *intangible materialism*—is to present the opposite of Pansy's narrative experience, the possibility of *nonaffective pain*. Such pain is also, in an odd way, the opposite of Teresa's and even the Hindu pilgrims' experience as well. This possibility pursues another kind of double articulation of suffering and pain—the fact, as Saussure has said, that in language nothing (a felt absence) can be taken up as a sign—a little different from Stafford's juxtaposition of Pansy and Teresa. In fact, nonaffective pain reduces Saussure's double articulation to the binary that is usually described as distinct from Peirce's tertiary semiotics: either the double articulation of meaning versus fact (the common mistake about Saussure) or the double articulation of meaning versus pattern apprehension (the other mistake within Saussure that seemingly prevents any inclusion of reference in his semiology). In the mid-twentieth century, the relatively widespread performances of lobotomies—Pansy's greatest fear—resulted, sometimes, in patients' "disconnection" from physical pain. "The lobotomy experience," Vertosick writes,

> taught neuroscientists that it's possible to dissociate the physical sensation of pain from the emotional perception of pain as unpleasant. Lobotomized patients still felt pain; they just didn't let it bother them anymore. Pharmacologists soon figured out ways to achieve this effect temporarily with drugs. Under dissociative anesthesia, the patient may be awake and receiving no analgesia, yet the sensations he or she feels aren't perceived as unpleasant. (2000, 214–15; see also Morris 1993, 143, and notes)

Pain without affect is not part of personhood: it has no narrative, no future, no past, and it occasions no suffering; it is neither iconic nor, in David Chalmers's conception, "conscious experience" at all. (In Chalmers—and the philosophical tradition he works in—experience without affect is consistently referred to as "zombie" experience.) Vertosick continues that "when adverted after the event, these drugs can also render a patient amnesic, with no recollection of the painful event at all. . . . The thought of having open-heart surgery while fully awake and totally paralyzed must rank as one of the most awful images the average intellect can conjure. Nevertheless," he says, "with the appropriate amnesic agent, we wouldn't remember any of it, so why should it matter?" (215). It does not matter—though there are cases on record of patients awaking, paralyzed in the midst of open-heart surgery, where it matters, after the fact, in their lawyer's narrative lawsuits—it does not matter because affect is a function of narrative, memory, anticipation, and subjectivity, and thus it is tightly connected to suffering, and these drugs allow pain without concern and create truly "meaningless" pain with its "unruly fragments, silences," its "shameful, painful, prelingual limitations" Charon and Spiegel describe. Thus, it is no accident, I believe, that Stafford resorts to the double articulation of premodern theological metaphors and Pansy's physiological pain. As Vertosick notes, the reason "why so many chronic pain patients descend into deep despair [is that] their pain is devoid of meaning. They have a suffering that's without purpose, a pain both sterile and pointless. They have the pain of hell—utterly useless—while [someone like the mountain climber Sir Edmund] Hillary's pain was of purgatory, a physical price to be paid for metaphysical reward [such as conquering Everest]" (176). The possibility of physiologically "remembered" but pointless pain, what I might call the "behavioristic" pain Kandel and Milner describe, with its erasure of meaning, purpose, and personhood, is the source of suffering and, theologically speaking, Hell itself. Its opposite, as any semiotician will tell you, is the purgatory of meaning—disorder brewing out of order—that can be articulated against the silences of mechanistic science or discovered within the seeming purport of evolutionary design, as we learn in the memories of the pain that we suffer.

Materialist Literature

The nonaffective pain I have portrayed is a striking example of reductive materialism: it is an example of matter conceived "as particulars, isolated from their contexts and immune from assumptions (or biases) implied by words like 'theory,' 'hypothesis,' and 'conjecture'" that Mary Poovey described (1998, 1). But I argue in this chapter—and throughout this book—that we can entertain a wider sense of matter and materialism, that if the *processes* of natural selection (as opposed to the *particulars* Poovey mentions) can be understood as material, then so can the *phenomenologies* of semiotics. I have argued that facts can be understood as something other than brute and that neither experience nor meaning is as "immediate" and "intuitive" as it seems, so that the phenomenologies of experience and meaning, even when they gather up absence and nothingness, can be understood in the negative science of semiotics. Such negativity is most clear in the phenomena of pain, which seems as meaningless as a simple fact, a more complicated fact to be resisted, and a phantom—or negative—fact to be transformed. In these ways, I suggest, pain both instantiates and destroys subjectivity. These three characteristics of the fact of pain—iconic, indexical, symbolic—characterize discourse and, I have been both arguing and, in my examples, indicating, they are most easily apprehensible in what is called literature.

In the Introduction, I offered a global sense of literature as privileged forms of verbal discourse and patterned uses of linguistic expression that go beyond the pragmatic function of communication and that have been a part of all organized human societies. One of the several definitions I presented described the privileged forms of literature as the linguistic embodiment of the sacred and mystical within the social formations of discourse. These privileged forms are affective, narrative, and create the "effect"—the "phantom"—of a presiding subject of experience. They do so by taking up the "modes of being" Peirce describes, iconic sensuousness, an indexical situation in the biological and social world, and symbolic future-oriented comprehension. In the privileged texts of literature, these modes of being are most pronounced in both their unity and their distinction. The most pronounced instance of literature conceived in this way, I suggest, might be found in the sacred pain that Glucklich describes in religious experience. Thus, it is no accident, I think, that in our Western tradition, the study of literature literally begins with Aristotle's poetics of tragedy, in which, by means of a *catharsis* conditioned by the very fact of theory of mind, audiences can experience both the suffering of represented characters and the kind of "cleansing" that sometimes accompanies pain.[35] But surely the *power* of language can be seen in the materiality of our seemingly unmediated responses to faces, to curses, to engagements with others as well. In literary texts—in Eliot's *Ash Wednesday,* Keats's "This Living Hand," Stafford's "Interior Castle," or even Bowen's *The Heat of the Day*—icon and index call attention to themselves not only as contributing to symbol but as taking on lives of their own. Such lives are "their own," I suggest, because of they are nested in levels of understanding that—like the "notations" of semiotics—shape and condition material experience and scientific knowledge. Semiotics and science are connected, and throughout this book, I have described the levels of understanding of physical, biological, and informational sciences in relation to Peirce's semiotic categories to suggest both the delineation of a materialist sense of language and literature and the ways that literature achieves its powers that might help us comprehend the interpretative work of science in our time.

Notes

Introduction

1. In *Modernism and Time*, I argue that in his accounts of quantum physics, Heisenberg replaces notions of simultaneity with the idea—based on the Copenhagen interpretation—of alternation. The "conception of complementarity," according to Heisenberg, encourages physicists "to apply alternatively different classical concepts which would lead to contradictions if used simultaneously" (1958, 179). Alternation, like Russell's periodicity (discussed in chapter 1), makes temporality a constituent element in any sense of materialism (see Schleifer 2000b, ch. 5, esp. 186–87).

2. I examine such a procedure at length in *Analogical Thinking* (2000a, esp. ch. 1).

3. In *The Disorder of Things*, John Dupré argues that the category of an "essence" or a "natural kind" both for physical objects and biological creatures can be delineated "only in relation to some specification of the goal underlying the intent to classify the object" (1993, 5), what I describe throughout *Intangible Materialism* as *notation* and *attention*, the very "noticings" of literature: "The idea that things belong to unambiguously discoverable natural kinds," Dupré writes, "is intimately connected with the commitment to essentialism . . . [which posits] that what makes a thing a member of a particular natural kind is that it possesses a certain essential property, a property both necessary and sufficient for a thing to belong to that kind" (1993, 6). Dupré rejects this idea in favor of the "pluralism" of something like Wittgenstein's "family resemblance concept" (1993, 10) based on "the underlying ontological complexity of the world, the disorder of things" (7). In a similar fashion, following Wittgenstein and Kuhn more explicitly, Andersen, Barker, and Chen (2006) describe "family resemblance" classifications in astronomy and biology. Together, these studies "reformulate" (to use Poovey's term) understandings of wholeness in terms of ontology and

epistemology (cognitive science). *Intangible Materialism* attempts to do this in terms of semiotics as well.

4. In *The Structure of Evolutionary Theory,* Gould offers an extended example of this difference. "The appearance of a 'face' on a large mesa of the surface of Mars—an actual case by the way, often invoked by fringe enthusiasts of extraterrestrial intelligence—bears no . . . conceptual homology to faces of animals on earth. We label the similarity in pattern as accidentally analogous—even though the perceived likeness can teach us something about the innate preferences in our neural wiring for reading all simple patterns in this configuration . . . as faces. (An actual face and the accidental set of holes on the mesa top may stimulate the same pathway in our brain, but the two patterns cannot be deemed causally similar in their own generation—that is, *as faces*)" (2002, 928). In chapter 2, I examine this phenomenon of apprehending faces, which Gould suggests provides an analogy of *function.* In chapter 3, I argue that there is a material *homological* relationship between automatic calling phenomena in mammals, and especially in primates, which are revealed in Tourette syndrome, and poetry (see ch. 3, note 6).

1. Intangible Materialism

1. In this catalog, Chalmers presents physical, biological, and cognitive examples of reductionism, the three categories I examine in chapter 2 and throughout this book. (John Dupré [1993] also follows these categories in his philosophical critique of reductionism.) In another delineation of reductionism, George Levine notes that "in the end, not only would human behavior be explicable by biology, but biology would be explicable by chemistry, and chemistry explicable by particle physics. . . . Reductionism," he concludes, "gives causal priority to one level of explanation over all others" (2006, 54). Edward O. Wilson offers a strong version of such reductionism throughout his career and especially in his book *Consilience,* which I examine closely in chapter 2. In fact, Levine describes Wilson's "strong reductionism" (125) and calls him—along with Richard Dawkins, Daniel Dennett, and Steven Pinker—a "biological determinist" (107), offering a sweeping survey, focused on Wilson, of the contemporary debate over reductionism. Dawkins and Dennett pursue sociobiology—a scientific program Wilson helped initiate early in his career (1975)—while Pinker pursues evolutionary psychology; in *Consilience,* Wilson subscribes to a basically physicalist reductionism. These three categories—physicalism, sociobiology, and evolutionary psychology—also describe the categories treated in *Intangible Materialism.*

"Much current resistance to reductionism," Levine argues, "is based on the argument . . . that reductionism, in assuming that all complex structures are built up out of smaller and less complicated ones, tends to ignore the problem of the relations *among* the smaller structures; it fails, then, to address the problem of how the relationships among the basic phenomena actually change the way those phenomena behave" (2006, 53). In his examination of consciousness, Chalmers does not participation is this kind of resistence to reductionism; rather, he posits a "naturalistic dualism." Outside of this metaphysics, as far as I can determine, Chalmers remains sympathetic to physicalist reductionism.

2. In specific terms, Chalmers notes that "reductive explanation is not the be-all and end-all of explanation. There are many other sorts of explanation, some of which may shed more light on a phenomenon than might a reductive explanation in a given instance. There are *historical* explanations, for instance, explaining the genesis of a phenomenon such as life, where a reductive explanation gives only a synchronic account of how living systems function. There are also all sorts of *high-level* explanations, such as the explanation of aspects of behavior in terms of beliefs and desires. Even though this behavior might in principle be explainable reductively, a high-level explanation is often more comprehensible and enlightening" (1996, 43). More generally, he argues throughout *The Conscious Mind* that consciousness is not simply susceptible to other explanations and that the *mechanisms* of materialism cannot explain or account for consciousness (conceived as the experience and apprehension of qualities) in the ways they can be applied to heat or earthquakes or life itself—because, he maintains, in the case of consciousness there is more to be explained than structure and function.

3. Chalmers uses the term *physicalism* as well, and uses it interchangeably with *materialism*. If by physicalism he means the necessarily mechanical understanding of explanation he describes in terms of reductionism, then perhaps it would be more useful to define materialism more widely, to encompass every kind of nonsupernatural explanation, not just reductive explanations in terms of structure and function.

If the distinction were made this way, then Chalmers would probably be a materialist but not a physicalist. My colleague James Hawthorne describes Chalmers's view this way: "It's a bit hard to say what Chalmers 'is' with regard to being a physicalist/materialist," he writes, "because although he suggests various possible solutions to the problem of the basis for experience, he is open to various alternatives. His bottom line is that the kinds of physical accounts we give for all other physical phenomena (via the functioning of physical mechanisms and processes) doesn't work as an account of phenomenal experience. So he is a dualist in that he thinks we need a wholly different kind of account here. Now, he is in principle willing to entertain any sort of account that could be reasonably worked out, and that would in fact work in accounting for experience. But his bet is that it will 'not' be some sort of 'Cartesian dualism,' based on non-physical spirits interacting with the brain somehow. His bet is that it will be some sort of materialist account. But he is 'dualistic' in the sense that he doesn't think it possible to give an account of phenomenal experience in the 'usual physicalist way'—i.e. in terms of the functional states of physical mechanisms and processes. . . . It's not that he thinks there is non-physical stuff. Rather, he thinks that 'physical processes and mechanisms,' as usually understood, can't do it" (personal correspondence). Chalmers, I think, is focusing on what neurologists have called "core" or "primary" consciousness, which I discuss in chapter 4 (see especially note 3), which lends itself to the material analyses and explanations of natural selection. In addition, in chapter 2, I pursue a nonphysicalist but *material* account of homological "nestings" that obviate the opposition between the physical and the phenomenal understandings by suggesting that phenomenal experience *emerges* from—and in doing so is not simply dependent on—"usual physicalist" explanations.

4. Discussing quantum physics and the role of the observer, John Wheeler makes a similar point: "Wigner speaks of the elementary quantum phenomenon as not really having happened unless it enters the consciousness of an observer. I would rather say that the phenomenon may have just happened but may not have been put to use. And it's not enough for just one observer to put it to use—you need a community" (1986, 63). In an analogous fashion, John Casti, discussing the mathematics of complex systems, asserts that "meaning [is] part of a cultural cognitive map" (1995, 273). Earlier, he notes that "meaning is bound up with the whole process of communication and doesn't reside in just one or another aspect of it. As a result, the complexity of a political structure, a national economy or an immune system cannot be regarded as simply a property of that system taken in isolation. Rather, whatever complexity such systems have is a joint property of the system *and* its interaction with another system, most often an observer and/or controller" (269). In *Analogical Thinking,* I examine the ancient conception of "theory" as public and communal witnessing rather than a private act. The *theoros* in Greek society were "certain individuals . . . [designated] to act as legates on certain formal occasions. . . . These individuals . . . collectively constituted a *theoria*" (Godzich 1986, vii). "The social mobilization of theory defined in this way," I argue, "*institutes* the private self-evidences of sight . . . as public discourses of knowledge, so that witnessing expands its significance, analogically, from the act of seeing to the act of telling" (2000a, 98).

5. Claude Lévi-Strauss, in many ways Greimas's tutor, describes it this way: "I have tried to transcend the contrast between the tangible and the intelligible by operating from the outset at the sign level. The function of signs, is, precisely, to express the one by means of the other. Even when very restricted in number, they lend themselves to rigorously organized combinations which can translate even the finest shades of the whole range of sense experience. We can thus hope to reach a plane where logical properties, as attributes of things, will be manifested as directly as flavors or perfumes; perfumes are

unmistakably identifiable, yet we know that they result from combinations of elements which, if subjected to a different selection and organization, would have created awareness of a different perfume" (1975, 14).

6. In their famous feud—which can be understood in terms of these two conceptions of "fact"—Erwin Schrödinger even said that the impossibility of visualizing Werner Heisenberg's mathematical description of quantum mechanics was "disgusting, even repugnant" (cited in Crease and Mann 1986, 57). In *Complexification,* John Casti pursues what he calls "a theory of models" (1995, 278) through his mathematical exploration of "the science of surprise."

7. Oliver Sacks makes a similar point. "Language and thought, for us," he writes, "are always personal—our utterances express ourselves, as does our inner speech. Language often feels to us, therefore, like an effusion, a sort of spontaneous transmission of self. It does not occur to us at first that it must have a *structure,* a structure of an immensely intricate and formal kind. We are unconscious of this structure; we do not see it, any more than we see the tissues, the organs, the architectural make-up of our own bodies. But the enormous, unique freedom of language would not be possible without the most extreme grammatical constraints" (1989, 74–75).

8. In *Modernism and Time,* I examine what I call the "non-transcendental disembodiments of solid goods, experiences, and even understanding" (2000b, 122) at the turn of the twentieth century. As in Bertrand Russell's work, the larger argument in that book focuses on the *temporalization* of experience and phenomena. Virtually all of the literary works discussed in the present book are twentieth-century works, though I believe earlier examples (such as the writings of Saint Teresa, which are also discussed) could easily be substituted, since the experience of living in and through the twentieth century conditions our reading more generally.

9. For a thorough discussion of this phenomenon, see Arrighi (1994); the "long" twentieth century he describes stretches back to the fifteenth century. Fischer (1996) also examines the long cyclical history of the replacement of industrial capital by finance capital in non-Marxist terms.

10. This is very different from Peirce's indexical referentiality that, like information, involves the subject in an act of attention which is part and parcel of indexical reference.

11. In *Analogical Thinking,* I elaborate on the force of analogy in meaning and understanding; see especially chapter 4, which discusses this passage from Greimas and Courtés. Earlier in the book, I note that Roman Jakobson specifically defines "meaning" as the translation of one set of signs into another set of signs (2000a, 35), a notion that aligns itself with Casti's definition of meaning as "part of a cultural cognitive map" (1995, 273).

12. In *Analogical Thinking,* I argue that "the conjunction of syntax and semantics—both their (formal) opposition and their (temporal) combinations—distinguishes the disciplines of humanistic study from the

nomological sciences. Such a conjunction creates the possibility of analogical thinking, which is neither haphazardly arbitrary nor logically formal" (2000a, 109).

In her study of human gestures that surveys contemporary work in experimental psychology, Susan Goldin-Meadow argues that gestures also function like such notation: "I argued earlier that gesture reflects a speaker's thought, often offering a novel view of what's on the speaker's mind. But perhaps gesturing does more than reflect thoughts; perhaps it also plays a role in shaping them. We know, for example that writing a problem down often makes it easier to solve the problem, either by reducing demands on memory or by providing a new perspective" (2003, 145); and she further argues—based upon experimental data—that gestures similarly "increase available cognitive resources" (158). These three examples of notation—Russell's example from physical chemistry, Goldin-Meadow's from social science, and Greimas's and Courtés's from semiotics—are arrayed in the categories I examine in chapter 2.

13. In *The Conscious Mind,* Chalmers concludes his study with a speculative description of consciousness based on Claude Shannon's formal definition of information. Chalmers almost necessarily follows Shannon's abstract mathematization of information because he is working on the level of logic rather than semantics; he is contrasting the abstract formalism of logic with phenomenologically rich semantics, even while he does not analyze, but only asserts (intuits) the "meanings" of experience. That is, he fully separates the two definitions of "interpretation" I presented earlier, creating, as he says, a "natural dualism."

14. In *Analogical Thinking,* I note that this reconceptualization—like the "reformulation" of fact Poovey describes—repeats "the great conceptual leap of Saussurean linguistics . . . , its self-conception, in Greimas's words, as 'a linguistics of perception and not of expression'" (2000a, 88). The rest of this paragraph paraphrases the argument I make in *Analogical Thinking.*

15. Prigogine and Stengers describe such a "universalized" and transcendental position for understanding as "a divine point of view from which the whole reality is visible" as impossible, since "no single theoretical language articulating the variables to which a well-defined value can be attributed can exhaust the physical content of a system. Various possible languages and points of view about the system may be complementary. They all deal with the same reality, but it is impossible to reduce them to one single description" (1984, 225; Dupré makes the same point in arguing for an anti-reductionist "pluralism [that] precludes the privileging of any particular level" or point of view [1993, 101]). That Prigogine and Stengers talk about vocabulary, just as I use narrative theory for my vocabulary—the "sender" and "receiver" found in Greimas's narratology (see ch. 5, note 22)—underlines the fact that the method and position of notation are not

irrelevant in comprehending material reality. *Notation* is another term for *observation* and the *observer* in discussions of quantum physics.

16. See the interviews collected by Davies and Brown (1986a). Especially apposite are the interviews with John Wheeler and David Bohm. Wheeler explicitly discusses "atoms of information" (1986, 66) and the ways in which information is "meaning" and is a function not only of individual but of communal "perception" (62, 63). Bohm describes—in a manner that might explicitly counter Chalmers—the ways that "life, mind and inanimate matter all have a similar structure" (1986, 123) in terms of information. These categories of matter, life, and mind correspond to my focus on physics, biology, and semiotics.

17. Tom Siegfried describes the "holographic principle" in relation to understandings of black holes in terms of information theory such that "the number of bits [of information] a black hole swallows is equal to the number of bits represented by its surface area. . . . Somehow, all the information contained in three-dimensional space can be represented by the information available on that space's two-dimensional boundary" (2000, 231).

18. Here is where Peirce's notion of the "interpretant" is so important: one way of understanding his concept of the interpretant is to recognize that what it does is *erase* information. Peirce's third category, symbol—in his catalog of icon, index, and symbol—offers *unmotivated* correspondences, what Saussure describes as the arbitrary nature of the sign. Saussure calls this a "word image," in an oxymoron that Peirce avoids with a purely unvisualizable neologism, "interpretant." A symbol, he writes, is "a sign which would lose the character which renders it a sign if there were no interpretant" (1931–35, 2:304). Sign and object necessarily give rise to a third element, the interpretant that, as another sign, is given to thought: "A sign," Peirce says, ". . . addresses somebody, that is, creates in the mind of that person an equivalent sign, or perhaps a more developed sign. The sign which it creates I call the *interpretant* of the first sign" (2:228). The interpretant, as John Sheriff notes, is "the thought which interprets" the sign–object relationship, and as such it is a necessary part of signification (Sheriff 1994, 34–35). Moreover, as a sign in its own right and the object of a sign as well, it *erases* as well as delineates information insofar as it replaces the sign of the preceding sign–object relationship. In this way, it lends itself to what Umberto Eco calls "unlimited semiosis" (1976, 69). Yet, like the "matter" or "things" Bohr and Heisenberg describe, it is, in Peirce's words, "altogether something virtual" (1931–35, 5:289). In this discussion, I am following Sheriff. See also Chandler 1994–2008 and Robert Scholes's chart situating the interpretant in relation to early twentieth-century philosophy (1985, 205; I reproduce this chart in chapter 4, note 7).

19. Event, I suspect, like the interpretant, is graspable yet unvisualizable: the apprehension of a collection of happenings as a cohesive event—the "meaningful whole" I discuss in a moment—itself might be a pure form of

the interpretant. For discussions of event in Foucault and literary narrative, see *Analogical Thinking* (2000a, 23, 173–76). For another discussion of material event, see Derrida 2001, 319.

20. The most famous example of what I call "dislocation," and what Wheeler describes as the absence of localization, arises in quantum theory in the form of "quantum nonlocality." Albert Einstein realized early on that quantum mechanics seems to imply some sort of nonlocal interaction or connectedness between distant parts of quantum systems. Together with two colleagues, Boris Podolsky and Nathan Rosen (thus EPR), he illustrated this implication with the now famous EPR thought experiment (Einstein, Podolsky, and Rosen 1935). This thought experiment shows that the commonsense "principle of locality," namely, that an object is influenced only by its immediate surroundings, is violated *unless* quantum particles have *definite* positions and momenta at all times (whether measured or not), in violation of the quantum "uncertainty principle." EPR argue that quantum theory is thus an "incomplete" theory and needs to be extended to a better theory that assigns particles precise positions and momenta (sometimes called hidden variables) at all times. In the early 1960s, John Bell proved a theorem that shows that what EPR suggested is not possible—that no physical theory of local hidden variables can ever reproduce all the predictions of quantum mechanics (1964). Bell's theorem applies to situations very similar to those employed in the EPR thought experiment. In 1982, Alain Aspect turned the thought experiment into an actual experiment and verified that quantum theory makes correct predictions in the situations to which Bell's theorem applies, thus showing that there really are "nonlocal connections" between distant parts of quantum systems, substantiating what I would like to describe as this "intangible" phenomenon—the *indefiniteness* of material location in the quantum world. I thank James Hawthorne for his contribution to this note. For a thorough—if somewhat technical—treatment of Bell's theorem, see Hawthorne and Silberstein 1995.

21. Also similar is Derrida's late articulation of "materiality without matter" (2001, 350): the word *materiality,* he writes, "brings with it a high essentializing risk where it should exclude, in its interpretation, any semantic implication of matter, of substratum or instance called 'matter' and any reference to some content named matter; it risks thereby meaning only 'effect of matter' without matter" (353).

22. This and the following two paragraphs are based on my discussion of Saussure in *Rhetoric and Death* (1990, 69–71).

23. Most handbooks describe Peirce's icon as, in the words of Daniel Chandler, "a mode in which the signifier is perceived as *resembling* or imitating the signified . . . (recognizably looking, sounding, feeling, tasting or smelling like it)—being similar in possessing some of its qualities" (Chandler 1994–2008). Chandler further describes the index as a signifier that "is *directly connected* in some way (physically or causally) to the signified" and

the symbol as what Peirce called "a conventional sign, or one depending upon habit (acquired or inborn)" (Chandler citing Peirce 1931–35, 2:297). While these definitions are accurate, they de-emphasize the *materiality* of meaning. In fact, in chapter 4, I make clear that the index is a function of our index fingers in a chapter devoted to the evolution of hands and their function in poetry. Such materiality is embedded in Peirce's conception of the sign and semiotics, particularly in his category of icon.

24. Oliver Sacks recently described the neurological breakdown of the simultaneous participation in these modalities—the technical term is *simultagnosia*—in relation to a composer who, after brain injury, could not integrate the sounds of musical pieces. In a note, Sacks says that "something analogous to a transient simultagnosia may occur with intoxication from cannabis or hallucinogens. One may find oneself in a kaleidoscope of intense sensations, with isolated colors, shapes, smells, sounds, textures, and tastes standing out with startling distinctness, their connections with each other diminished or lost" (2007, 115). Chalmers argues that he can imagine a "possible world" of zombies, people who act without the "felt experience" of consciousness. I would think the dysfunction of simultagnosia is another "possible" world (different from our own) of the isolated experience of qualities that Chalmers claims for *our* world—in which, as Peirce suggests, experiences of consciousness "of whatever is at any time before the mind in any way" are mediated by the simultaneous apprehension of differing modalities of classification. Such "classes," as Dupré argues (1993) and Peirce suggests, are not simply epistemological but different *kinds* of facts.

25. For a thoughtful discussion of accounting for phenomena that are not—and often *cannot* be—observed, see Bogan and Woodward (1988), and especially the distinction they make between data and phenomena, the distinction between "what is uncontroversially observable (data)" and "what theories explain (phenomena or facts about phenomena)" (314). Instances of phenomena, they note, "can occur in a wide variety of different instances or contexts. This, in turn, is closely connected with the fact that the occurrence of these instances is (or is plausibly thought to be) the result of the interaction of some manageably small number of some causal factors, instances of which can themselves occur in a variety of different kinds of situations. . . . Data are, as we shall say, idiosyncratic to particular experimental contexts, and typically cannot occur outside of those contexts. . . . Phenomena, by contrast, are not idiosyncratic to particular experimental contexts. We expect phenomena to have stable, repeatable characteristics which will be detectable by means of a variety of different procedures, which may yield quite different kinds of data" (317). This distinction is implicit in Peirce's conception of abduction, and especially in his conception of the "character" or "characteristics" of phenomena upon which abduction focuses. It also recapitulates the two interpretations of fact that Poovey describes.

26. In *Criticism and Culture,* R. C. Davis and I examine "the negation of [Saussure's] critique of linguistics as he found it" as "best seen not as a different 'moment' in his reasoned critique but as its constant 'other.' Throughout his career, as he developed and taught many of the seminal concepts that have governed the way we understand language and, indeed, other phenomena in the twentieth century, he worked concurrently on a lifelong study of possible anagrams inscribed silently in late Latin poetry. More specifically, as he records in one of his notebooks, the presence of coded *parts* of proper names in poetry might, in fact, constitute the power of poetry: 'the reason [for the presence of anagrams in Latin lyrics],' he writes, '*might have originated* in the religious idea that an invocation, prayer, or hymn would have power only if the syllables of the divine name were worked into the text.' In other words, Saussure's life work to discover the 'sufficient reason' for the existence of *meaningful* language exists concurrently with his life work to discover patterns in the more or less meaningless sound patterns of language" (1991, 33). In chapter 3, I examine the relation of meaningless sound patterns to the power of poetry in the concrete, material example of Tourette syndrome. Levinas, like Saussure, also examines the possibilities of the acknowledgment of strangeness without positing the kind of dualism Chalmers suggests.

27. "Those of us," Gould writes in *The Structure of Evolutionary Theory,* "who practice the science of reconstructing specific events and unravelling temporal sequences have always fought a battle for appropriate status and respect, no less so today than in Darwin's time" (2002, 102).

28. Walter Benjamin brilliantly analyzes the concept of wholeness under the category of "constellation." For a discussion of Benjamin's thought on this matter—including its historical occurrence in the period "after about 1870" Poovey describes—see my *Modernism and Time* (2000b, ch. 3).

29. For an examination of tragedy in the context of "the cerebral basis of language, art, and belief" (the book's subtitle), see Trimble (2007, esp. 191–97). A sense of the meaningful whole of narrative is particularly prominent in detective fiction, whose explicit goal is to present a fact or a situation that allows the many events of the narrative to fall into place—to be *explained*—as elements within a larger, "whole" pattern of action, of cause and effect. This is why the essays collected in *The Sign of the Three,* edited by Umberto Eco and Thomas Sebeok, consistently focus on the relationship between Peirce's semiotics—and especially his sense of abduction—and the narratives of Sherlock Holmes. In a technical philosophical analysis of abduction, Ilkka Niiniluoto also notes that "a historically interesting application of abduction as a heuristic method can be found in classical detective stories" (1999, S440). See also Hunter 1991 and Schleifer and Vannatta 2006 for practical use of detective narrative in medicine.

30. This catalog is suggested in many post-Peircean studies of abduction and explanation. For instance, in their discussion of the nature of

explanation—and especially *scientific* explanation—in "Saving the Phenomena," Bogan and Woodward are careful to describe explanation as entailing "detailed causal stories" as well as the "classical deductive-nomological model" of mathematics (1988, 324). Their example of a story or narrative explanation is an account of the cause of Parkinson's disease, and they claim that the deductive model often entails "idealizations and approximations" (324n). In an earlier discussion of "inference to the best explanation," Gilbert Harman argues in his description of the logic of abduction that "enumerative induction" is a special case of abduction: "inference to the best explanation" (1965, 88).

31. This assertion distinguishes Peirce from the positivists. "For classical positivism," Dupré writes, "reductionism refers to the attempt to ground knowledge entirely in the simplest possible observation statements, so as to eliminate any dubious inferences beyond the observable" (1993, 8). Darwin's historical science is also distinguished from positivism in this way. Thus, Gould argues that there is an inability to "observe" history—not only because of the imperfect record of phyletic history but also because we cannot "'see' past causes directly, but only draw conclusions from preserved results of these causes" (2002, 59), since "past processes are, in principle, unobservable" (102; as we shall see, this is Peirce's notion as well). For this reason, Gould notes that, as Darwin wrote, "change of species cannot be directly proved" (cited in Gould 1986, 65).

2. The Faces of Consilience

1. See Elizabeth Ermarth's fine description of the advent of scientific "law" in the Enlightenment: "When time and space are conceived as homogeneous—that is, the same universally—then it becomes possible to chart both the differences and similarities in nature which give rise to those generalizations in science and art that we call laws. In formulating such laws no attempt is made to save the appearances. In fact, we might say that in reducing the welter of particulars to some abstract regularity, scientific and realistic generalizations represent an attempt to save the essences" (1983, 17–18). For extended discussions of the criteria for accuracy, simplicity, and generalizability as the criteria for (Enlightenment) science, see my *Analogical Thinking*. See also Schleifer, Davis, and Mergler 1992, 1–37. In a way, Chalmers's study, examined in chapter 1, attempts to save the appearances—phenomenology—in the face of nomological science.

2. Fritjof Capra quotes Descartes in a similar vein: "'All philosophy is like a tree,' wrote Descartes. 'The roots are metaphysics, the trunk is physics, and the branches are all the other sciences'" (1997, 12).

3. See also P. C. W. Davies's and J. R. Brown's more general description: "Today," they write, "many scientists . . . prefer to regard the brain as a highly complex, but otherwise unmysterious electrochemical machine,

subject to the laws of physics in the same way as any other machine. The internal states of the brain ought therefore to be entirely determined by its past states plus the effects of any incoming sense data. Similarly, the output signals from the brain, which control what we like to call 'behaviour,' are equally fully determined by the internal state of the brain at the time" (1986b, 32). A key term here is *effects* insofar as the effects of incoming data and the effects described as a "state" function within the simplicity of a mechanical cause-and-effect operation. Chalmers takes the mechanistic—machinelike—nature of the brain to be only half of a "natural dualism," which also posits nonmechanistic phenomenologies of experience (1996).

4. A significant limitation of Steven Weinberg's "dream" of a final theory encompassing the "fundamental laws of nature" is the ease with which he dismisses history as purely "accidental" (1992, 37). He discusses "the intrusion of historical accidents" in terms of "the buzzword 'emergence'" (39), and offers the example of thermodynamics as "more like a mode of reasoning than a body of universal physical law" (41). Percy Bridgman has observed that the laws of thermodynamics have "something more palpably verbal about them—they smell more of their human origins" (1941, 3; cited in Campbell 1984, 278); Douglas Hofstadter uses thermodynamics as a particularly clear example of levels of understanding.

5. Such connections across classes of facts can be seen in the pleasure Gould takes in the fact that "ironically, Whewell, a conservative churchman, later banned Darwin's *Origin* from the library of Trinity College, Cambridge, where he was master. What greater blow than the proper use of one's own arguments in an alien context" (1986, 65). In chapter 4, I pursue Gould's sense of history further.

6. Fraser's conception of "nested" levels is related to the concept of "irreversibility" nicely presented by Ilya Prigogine and Isabelle Stengers in *Order Out of Chaos*. They contrast the irreversibility of thermodynamics (and, to a lesser extent, molecular biology) with the reversibility of dynamics and classical physics. They argue, as I suggest here, that the opposition between physics and biology does not require, at some level, dualism in comprehending a materialist understanding of nature. "As we have emphasized repeatedly," they argue, "there exist in nature systems that behave reversibly and that may be fully described by the laws of classical or quantum mechanics. But most systems of interest to us, including all chemical systems and therefore all biological systems, are time-oriented on the macroscopic level. Far from being an 'illusion,' this expresses a broken time-symmetry on the microscopic level. Irreversibility is either true on *all* levels or on none. It cannot emerge as if by a miracle, by going from one level to another" (1984, 285). Just as Fraser—and Wiener and MacKay—indicate the close connection between nesting and communication, so Prigogine and Stengers note that "irreversibility and communication are closely related" (295).

7. "I think you are still having some difficulty realizing the difference

in levels here," Hofstadter writes in one of the fugue dialogues of *Gödel, Escher, Bach.* "Just as you would never confuse an individual tree with a forest, so here you must not take an ant for the colony" (1979, 315). Holland also repeatedly describes the relationship between microscopic and macroscopic descriptions as that between the trees and the forest, and he does so to emphasize his constant argument that "*the whole is more than the sum of the parts.* . . . The interactions between the parts are nonlinear, so the overall behavior *cannot* be obtained by summing the behaviors of the isolated components. Said another way, there are regularities in system behavior that are not revealed by direct inspection of the laws satisfied by the components. These regularities both explain (parts of) the system's behavior and make possible activities, and controls, that are highly unlikely otherwise" (1998, 225).

8. Pressure, color, temperature—Descartes's and Locke's "secondary" qualities—are problems for mechanical positivism precisely because they focus upon the interaction of different levels of phenomena. As I noted in chapter 1, Norbert Wiener focuses on this when he contrasts Leibniz and Newton in terms of the precedence Leibniz gives to optics (1967, 26–27). Leibnizian optics, he argues, unlike mechanics, cannot easily situate the subject outside time because it is precisely concerned with information, messages, and communication. See *Modernism and Time* for a discussion of this opposition and its appropriateness in relation to Heisenberg's description of quantum physics (2000b, 203–7), some of which I discuss in chapter 1. Holland discusses levels of understanding in terms of the "appropriateness" of frameworks "that are well-tuned to the question(s) at hand" (1998, 185).

9. In *The Right Brain,* Robert Ornstein repeats Geertz's contention: "After the modern brain developed, toolmaking and the beginnings of society changed the nature of evolution itself. Instead of adapting to the external environment, human society *became* the environment for the most part, and our own evolution was more and more in our own hands" (1997, 34).

10. In philosophy, discussions of the mind–body problem organize themselves in terms of physics, biology, and mind. Semiotics, rather than mind, as the third category, allows for the integration of information theory into the three-leveled materialism I am describing. For overviews of this philosophical work, see Blitz (1992), Kim (1993), and Klee (1984) as well as Chalmers's (1996) thorough engagement with it.

11. Walter Moore, Schrödinger's biographer, also notes that in this book Schrödinger introduces "what was to become one of the most fundamental concepts in the new science of molecular biology: *the chromosome is a message written in code.* . . . A few earlier works had hinted at the idea of a code in the chromosomes, but Schrödinger was the first to state the concept in clear physical terms" (1992, 396). Among those whom the book influenced, Moore notes, was James Watson, who wrote, "From the moment

I read Schrödinger's *What Is Life?* I became polarized towards finding out the secret of the gene" (cited in Moore 1992, 403). "No doubt," Moorc concludes, "molecular biology would have developed without *What Is Life?* but it would have been at a slower pace and without some of its brightest stars. There is no other instance in the history of science in which a short semipopular book catalyzed the future development of a great field of research" (404).

12. Since 1943, advances in the understanding of evolution at the molecular level have substantiated Schrödinger's claim. In *Order Out of Chaos,* Prigogine and Stengers describe aspects of molecular biology that reinforce Schrödinger's insight. "Let us emphasize," they write, "the basic conceptual distinction between physics and chemistry. In classical physics we can at least conceive of reversible processes such as the motion of a frictionless pendulum. To neglect irreversible processes in dynamics always corresponds to an idealization, but, at least in some cases, it is a meaningful one. The situation in chemistry [and particularly the chemistry of molecular biology, as their examples assert] is quite different. Here the processes that define chemistry—chemical transformations characterized by reaction rates—are irreversible. For this reason chemistry cannot be reduced to the idealization that lies at the basis of classical or quantum mechanics, in which past and future play equivalent roles" (1984, 137). Later, they note "an initial element marking the difference between physics and biology. Biological systems *have a past.* Their constitutive molecules are the result of an evolution; they have been selected to take part in the autocatalytic mechanisms to generate very specific forms of organization processes" (153). For a thorough analysis of evolution on the level of molecules and the "irreversibility" of biological phenomena, see Johnjoe McFadden, *Quantum Evolution* (2000).

13. Such a sense of arrangement and organization takes its place among the "postmodern" facts that Mary Poovey describes as developing at the end of the nineteenth century. In *Modernism and Time,* I note that Bertrand Russell, describing contemporary mathematics in 1917, argued that "in former days, it was supposed (and philosophers are still apt to suppose) that quantity was the fundamental notion of mathematics. But nowadays, quantity is banished altogether, except from one little corner of Geometry, while order more and more reigns supreme" (1917, 87). Later in the twentieth century, and more generally, Walter Benjamin describes "conceptual constellations" (Schleifer 2000b, 163) as the best way of comprehending ideas.

14. Wheeler calls this quality "distinguishability," which is, he says, "a central point in the establishment of what we call knowledge or meaning" (1986, 62). This is, of course, Whewell's and Peirce's different "classes" of fact.

15. In *Analogical Thinking,* I examine Jacques Derrida's relation to linguistic science and the way that he takes up the categories of linguistic

"neutralization" and linguistic "marking" and "unmarking"—in which a functioning opposition ceases to work in particular contexts (e.g., the opposition of old versus young is not appropriate in "I am ten years old")—that is a particular instance of units of communication with the *same* dimensions being recognized as *not* equivalent (2000a, 57–67). "Deconstruction," as Derrida mobilizes it in his early work, I argue, displaces difference "in a kind of neutralization that is no neutralization at all but, rather, *negates* ('explodes') neutralization, 'resisting and disorganizing it,' as Derrida says, '*without ever* constituting a third term, without ever leaving room for a solution in the form of speculative dialectics" (2000a, 62). Derrida calls this "irreducibly nonsimple" (1982, 13, avoiding the simpler term *complex*) that, in important ways, results, as Greimas suggests, in the production of disorder from order.

16. Traditional materialism, Georges Bataille argues, is simply the obverse of idealism and participates in its assumptions. "Matter," he writes, "in fact, can only be defined as the *nonlogical difference* that represents in relation to the *economy* of the universe what *crime* represents in relation to the law" (1985, 129). For a discussion of Bataille's resistence to traditional materialism—which he claims is simply a variant of the same "metaphysical scaffolding" as idealism: "Two verbal entities are thus formed," he writes, "an abstract God (or simply the idea) and abstract matter" (45)—see Schleifer 2000b, 86–90.

17. A good example of this is the mathematical paradox of the sorites that Brian Rotman describes in his study of infinity. *Sorites* is a term derived from the Greek word for *heap,* and the paradox focuses on the point at which a collection of grains of sand "becomes" a heap or, in another example, subtractions of hair make one bald. "One cannot expect," Rotman argues, ". . . to point to individual numbers at which shifts in the structure of numbers occur any more than one can single out which individual grain's or hair's subtraction 'causes' a qualitative change. It is always a question (for both heaps and numbers) of zones and regions, of the boundaries of relations between global and local characteristics, of the emergence of novelty at a higher level than that at which particular numbers are cognized" (1993, 57–58). "Local" characteristics are particular grains of sand, and the presence or absence of a particular grain of sand makes no difference so that the category "heap" does not apply. The "global" characteristic of "heap" is, as in Peirce, a difference in *kind,* and this difference creates the very "levels" Rotman is describing. For Rotman's discussion of hierarchies of cognitive levels, see 1993, 107–8.

18. For a good discussion of such constraint, see Hayles's essay "Constrained Constructivism" (1991).

19. Paul de Man sets forth this problematic in his essay on Pascal (1996, esp. 58–59). See also Andrzej Warminski's powerful discussion of this essay,

and especially the functioning of zero, in which he claims (with de Man) that, unlike an indivisible "one"—that is "both a number and a nonnumber" (1996, 27)—zero "is radically not a number, absolutely heterogeneous to the order of number" (27; citing de Man 1996, 59) but only, in de Man's terms, "appears in the guise of a *one,* of a (some)thing" (59). Such absolute heterogeneity avoids the "ambiguity"—and undecidability—of zero that I am describing (and also the status of a material "thing" I discussed in chapter 1).

20. This and the following paragraph are based on correspondence with my colleague James Hawthorne. Conversations with Jim have vastly clarified my understanding, even if we do not always reach the same conclusions. He has helped me to understand more fully how the kind of work he and David Chalmers pursues can help all of us working in the humanities to achieve greater clarity and precision.

21. "Life," Johnjoe McFadden argues, "seems to emerge only at higher levels. It is, to use modern jargon, an 'emergent' phenomenon: one that cannot be entirely understood in terms of its parts. As a means to explaining life, the unrelenting reductionist approach is doomed to failure" (2000, 11). Near the end of his study, he claims that "the cell's ability to capture low-entropy states by performing internal quantum measurement is, I believe, fundamental to what it means to be alive" (256).

22. More specifically, Casti defines catastrophe theory as "an attempt to go beyond the confines of classical physics by providing a mathematical framework for describing these types of discontinuous processes" (namely, "price fluctuations on Wall Street, the breaking of waves on the beach and the outbreak of an infectious disease," again cultural, physical, and biological phenomena) (1995, 45). More technically, he notes that "catastrophe theory deals with the fixed points of families of [mathematical] functions. The catastrophes occur when, as we move in a continuous way through the family—usually by smoothly changing parameters describing the system—a stable fixed point of the family loses its stability" (55).

23. The most formal structure for such speculation that I know—one that enlists structure and function for the purposes of speculation—is Greimas's semiotic square that I described earlier in relation to the opposition of cold and hot. For an example of its speculative functioning—that combines algorithmic method and semantic unfolding—see *A. J. Greimas and the Nature of Meaning.* There, beginning with Lévi-Strauss's combination of agriculture versus war versus hunt (which Greimas takes up in *Structural Semantics*), I generate new combinations of meanings ("content") through ordered speculation that Greimas's methodical application of hierarchy and its negation provokes (1987, 30–33).

24. Jared Diamond also addresses this question and, more explicitly than Wilson, answers by switching his focus from the level of parts to the level of wholes (1992, 127) that I cited in the examination of "wholeness" in the Introduction.

3. Material Voices

1. For a remarkably detailed history describing the controversy of whether Tourette syndrome is an organic or a psychogenic illness, see Kushner 1999. Kushner thoroughly examines the major scientific studies of and cultural conflicts surrounding Tourette syndrome in the last century and concludes that "Tourette syndrome is an organic disturbance brought about by malfunctions connected with signaling in the basal ganglia" (192–93).

2. Chomskyan linguists limit language to these areas. Thus, Steven Pinker argues that "we can narrow down our search" for the brain centers of language "by throwing away half of the brain" (1994, 299). Other linguists suggest that language, like many other brain functions, is a distributed network that utilizes wide areas of the brain (see Lieberman 2000; Deacon 1998; Trimble 2007). But all agree that the cortex and neocortex are important, especially in more abstract forms of thinking: "functional magnetic resonance imaging (fMRI) data confirm the cognitive role of the dorsolateral-prefrontal-striatal circuit [the frontal lobe] in intact human subjects" (Lieberman 2000, 113; see also Deacon 1998, esp. ch. 10; Lieberman 1998, 70).

3. Howard Kushner identifies postencephalic tics and those of Tourette syndrome (1999, 66–71). L-DOPA stimulates the production of the biochemical neurotransmitter dopamine, and drugs that suppress the symptoms of Tourette syndrome, such as haloperidol, block dopamine.

4. It is notable that Gunn is describing in this poem the felt *coincidence* between semantic "sense"—the mysterious future-oriented "purport" of language that Louis Hjelmslev describes as the best definition of meaning (1961, 55)—and physical movement, a coincidence that Tourette syndrome, in its combinations of phonic and motor tics, clarifies by disrupting. Here, I need to acknowledge, and I hope to participate in, Sacks's remarkable and powerful sympathy and respect for people suffering from the "variety of involuntary and compulsive" motor and phonic tics, including people suffering from Tourette syndrome. In *Awakenings* and in later essays more directly focused on Tourette syndrome, Sacks, like W. H. Auden's father, Dr. G. A. Auden, does not always regard Tourette syndrome "as purely deleterious or destructive in nature. Less zealous to 'pathologize' than many of his colleagues," Sacks observes, "Dr Auden noted that some of those affected, especially children, might be 'awakened' into a genuine (if morbid) brilliance, into unexpected and unprecedented heights and depths" (1999, 17). This is certainly the phenomena he describes in the case histories of Tourette syndrome "Witty, Ticcy Ray" (1987), "A Surgeon's Life" (1995), and "Music and Tourette's Syndrome" (2007).

5. Lieberman argues, "What is offered here is a starting point, a theory that relates phenomena that are seemingly unrelated, such as why the pattern of deficits associated with the syndrome of Broca's aphasia involves

certain aspects of speech production, lexical access, and the comprehension of distinctions in meaning conveyed by syntax, why similar effects occur in Parkinson's disease, why children are able to learn language after massive cortical damage, why recovery from certain types of brain damage is problematic, and why aged people who speak slowly also have difficulty comprehending distinctions in meaning conveyed by complex syntax or long sentences" (2000, 16–17). Lieberman does not examine Tourette syndrome, yet it seems clear that his attempt to relate "phenomena that are seemingly unrelated" helps to establish the connection between the motor and phonic tics of Tourette syndrome. Neuroanatomical studies as well as Sacks's narrative histories argue for the connection between the complex symptoms of Tourette syndrome and the movement disorders (the motor dysfunction) of Parkinson disease (see Leckman, Riddle, and Cohen 1988, 104–5, as well as Sacks's repeated references to Tourette's in the latest revision of *Awakenings* [1999]). What is striking about Tourette syndrome is that its motor dysfunction is combined with phonic (and sometimes verbal) tics. Broca's area of the brain has also been shown to be an area of motor activity, and in the next chapter I touch upon some of the research that associates the motor activities of the hand and the motor activities of speech.

6. In a powerful close reading of Walter Kaufmann's translation of *The Birth of Tragedy,* Andrzej Warminski demonstrates that the "healing" project of semiology, as Deleuze describes it, should not be too sanguinely assumed. Rather than making Dionysian darkness visible, he argues, we can see in Nietzsche "more a stutter than a translation" from darkness to light, from noise to art: Kaufmann "could not make the intralinguistic transition (from German to English) precisely because he insisted on making the transition from music to images where the text itself did not, could not, make it—except as an impossible combination of words, a stutter . . . , a monstrous translation" (1987, xlv). His figure of stutter recalls Tourette syndrome. In this passage, he is substantiating Paul de Man's terrible observation that Nietzsche's "famous quotation ['Only as an *aesthetic phenomenon* is existence and the world forever *justified*'], twice repeated in *The Birth of Tragedy,* should not be taken too serenely, for it is an indictment of existence rather than a panegyric of art" (1979, 93; see also Derrida's late reading of de Man, esp. 2001, 350). Such a reading emphasizes the terrible condition of Tourette syndrome that takes up or sets forth the clarity of language as meaningless sound, dark materiality. In this chapter, I am not arguing for an analogy between Tourette syndrome and poetry but only for a material, *homological* relationship between primitive, automatic phonic signaling revealed in Tourette syndrome and its appearance in poetry. Stephen Jay Gould describes homology as "the well-established . . . concept of joint possession due to common ancestry" (2002, 601), and it is the "joint possession" of the *material* phenomenon of neurological responses that I am exploring here. For a discussion of the ways that Cole

Porter transforms lyrics to sound in his songs and seemingly unleashes "impersonal" desire, see my "Cole Porter and the Rhythms of Desire" (1999).

7. Frank Wilson also notes this phenomenon in his study of the human hand: "Because of the body's virtually limitless freedom to employ convenient combinations of muscle activity to achieve a desired movement, individual muscles of the body are subject to recruitment by the brain for assistance in an endless variety and range of movements. What begins as a specific behavioral demand—'Catch that ball!'—can usually be satisfied by any of a large set of alternative biomechanical solutions" (1999, 89). Deacon, in his richly detailed neurological study, also describes evolution as different from a simple logical hierarchy, noting that it is organized around "different ways of achieving the same goal" so that "neural distribution of language functions need not parallel a linguistic analysis of those same functions" (1998, 286). These descriptions—along with Lieberman's—underline alternatives to the logic of reductionism. They are versions of Gould's sense of "analogy" in that they are different "responses to common situations" (1986, 66).

8. As we have seen, in *Structural Semantics,* Greimas describes this as "the still very vague, yet necessary concept of the meaningful whole set forth by a message" (1983, 59). Sacks describes a similar phenomenal "signifying whole" in terms of the permanence of a biological "engram" that suggests that "one's *persona* is in no way 'localisable' in the classical sense, that it cannot be equated with any given 'centre,' 'system,' 'nexus,' etc. but only with the intricate totality of the whole organism" (1999, 239). In chapter 5, I examine this in conjunction with Ronald Melzack's discussion of the "neurosignature" of individuals (1995), habitual neurological patterns that create a sense of a "whole" self. For an extended discussion of Greimas's notion of "meaningful whole," see Vannatta, Schleifer, and Crow 2005, especially chapters 2 and 4.

9. One of the oddest symptoms of Tourette syndrome—one I return to later when I discuss the impulse to kiss other orphans that the narrator of Lethem's novel, *Motherless Brooklyn,* feels—is the urge to bite and lick things. A similar symptom (found in rhesus monkeys investigating "objects with their mouth instead of their hand") often accompanies Klüver-Bucy syndrome surgically induced in primates. This syndrome also includes "profound emotional disturbances" (Emery and Amaral 2000, 162; see also Aggleton and Young 2000, 108). In fact, the Klüver-Bucy syndrome is a significant part of the evidence of the role of the subcortical amygdala in emotion, including verbal signals in primate mating (Emery and Amaral 2000, 174).

10. In chapter 4, I mention that the experimental psychologist Susan Goldin-Meadow suggests that there is something "natural" about pointing toward something: "Even before young children produce pointing gestures to orient another's attention toward an object, they respond to others'

pointing gestures by directing their attention to the object indicated by the point" (2003, 91–92).

11. In his translation of Saussure's *Course in General Linguistics* (1983), Roy Harris renders *signifier* as "signal" rather than signifier. This is a problem—in a translation, I should add, that vastly improves the existing one—on two levels. First is the level of culture: for decades, *signifier* has been the common term, and one can hardly ignore a term's history and currency in making a translation. And second, on the level of sense: Saussure's *signifier* is precisely *not* a signal insofar as it exists in a structure of double articulation.

12. *Rime riche* describes rhyming in medieval and early modern poetry that rhymed homonyms such as *stair* and *stare* or *well* (adjective) and *well* (noun). In modern poets like Eliot and Yeats, such rhyming forgoes the semantics of "rhyming" different meanings and rhymes two signs as wholes, emphasizing their shared material sound. This is even clearer in music, such as Cole Porter's rhyming words with themselves that thus emphasizes sound over (or at least along with) sense. In an essay examining the relationship between music and lyrics in Cole Porter's songs, "'What Is This Thing Called Love?' Cole Porter and the Rhythms of Desire," I present a longer analysis of the example I use here from Yeats (1999, 17–19). I delineate—and indeed locate—the function of desire, which is arguably like Tourette syndrome and perhaps like music as well, situated between biological and cultural formations. There is a materiality to music at least analogous to the materiality of sounds and sound images. Mithen (2006) offers a strong history of these connections throughout his book.

13. Mithen (2006) offers an extended argument that partially refutes this assertion; Trimble (2007) does as well.

14. Deacon notes that "language evolved in a parallel, alongside calls and gestures, and dependent on them—indeed, language and many human nonlinguistic forms of communication probably co-evolved. . . . This is demonstrated by the fact that innate call and gesture systems, comparable to what are available to other primates, still exist side by side with language in us. Their complementarity with and distinction from language are exemplified by the fact that they are invariably produced by very different brain regions than are involved in speech production and language comprehension" (1998, 54). As I note in chapter 4, however, there are strong arguments (especially in Mithen, Donald, and Corballis) for the evolutionary relationship between gesture systems and language. Susan Goldin-Meadow (2003) describes in great detail the relationship between gesture systems and language.

15. For a fine discussion of the philosophical opposition between the "use" of a term in discourse and its "mention" when it is being discussed, metalinguistically, without functioning within discourse, see Jonathan Culler (1983, 28–29n11). While Culler argues for the deconstruction of this

opposition, I employ it here to distinguish between intentional use and unintentional mention. Later, however, I note the manner in which behaviors Sacks describes as "pseudo-actions" complicate the force of the Tourettic mention of obscenities (see note 21). And, of course, the place of intention—and especially *personal* intention that people like John Searle call upon (1969)—in relation to the *power* of language and poetry is precisely the problem of materialism that I describe in this book and that I focus on in chapter 5.

16. The *Diagnostic and Statistical Manual of Mental Disorders,* 3rd ed., rev., categorizes obsessive-compulsive behavior as "intentional": compulsion, it states, requires "representative, purposeful and intentional behavior that is performed according to certain rules or in a stereotyped fashion" (cited in Kushner 1999, 197). This definition has generated strong disagreements concerning whether obsessive-compulsive behavior is part of Tourette syndrome between those who contend that Tourette syndrome is *simply* organic and completely unintentional in its manifestations and those who contend it is psychogenic. Lieberman notes that "many neurological disturbances in humans (such as Parkinson's and Huntington's diseases, obsessive-compulsive behavior, depression) can be traced to disrupted basal ganglia circuits" (1998, 106). The ability to "suspend" or postpone motor and phonic tics in response to contexts (Kushner 1999, 7; see also Sacks 1987, 1995)—analogous to the ways we sometimes postpone sneezing—also creates the effect of blurring the opposition between intentional and unintentional behavior.

17. As we see in chapter 4, Robin Dunbar has made a sustained argument of the relationship between grooming and the evolution of language (1996).

18. In this regard, it might be useful to compare Eliot's description of poetry to Antonio Damasio's description of cognition in an experiment designed to elicit reasoned predictions from subjects, some of whom suffer from brain damage. "I suspect that before and beneath the conscious hunch there is a nonconscious process gradually formulating a prediction for the outcome of each move, and gradually telling the mindful player, at first softly but then ever louder, that punishment or reward is about to strike *if* a certain move is indeed carried out. In short, I doubt that it is a matter of only fully conscious process, or only fully nonconscious process. It seems to take both types of processing for the well-tempered decision-making brain to operate" (1994, 214).

19. Again, this comports with the structural linguistics in the context of which Greimas and Jakobson offer their analyses of poetry. As I have already mentioned, early in his career, Greimas describes the transformation of historical to formal or structural linguistics in the twentieth century as the creation of "a linguistics of perception and not of expression" (1963, 57; my translation). By this he means that linguistics turned to phenomenology and pursued the manner in which signification is apprehended rather than

(or along with) the ways it is generated. Certainly, the language of Tourette syndrome can be understood to be subject to a phenomenological poetics. In fact, I suggest that a neurological examination of some of the "impersonal energies" inhabiting Tourette syndrome particularly and language more generally lends itself nicely to a phenomenological materialism. As we shall see in the following chapters, Peirce far more urgently than Saussure insisted on such phenomenological materialism in his categories of index and icon. In fact, his third category, symbol, nicely corresponds to Greimas's "meaningful whole" that I have repeatedly mentioned here so that these chapters traverse Peirce's symbol, index, icon, moving almost backward against the usual presentation of his catalog of the aspects of signs.

20. For Greimas, such semantic categories, following the structures of Saussurean linguistics, are purely *relational:* in *Structural Semantics,* for instance, he "sketches" what he calls "the *semic system of spatiality*" (1983, 36) in terms of binary oppositions. (That is, he analyzes spatiality as consisting of dimensionality vs. nondimensionality; and dimensionality consisting, in turn, of horizontality vs. verticality; and horizontality consisting, in turn, of perspectivity vs. laterality; etc.) In a similar fashion, in an appendix to *Awakenings,* "Parkinsonian Space and Time," Sacks aligns Parkinsonian experiences of time and space with Leibniz's relative conceptions of time and space as opposed to Newton's absolute conceptions, calling the latter "convenient (or conventional) constructions or 'models'" (1999, 339). Again, similarly, Darwinian accounts of natural selection describe adaptive strategies as "proximate," and, like the Saussurean notion of the "arbitrary nature of the sign" or Sacks's notion of the conventionality of space-experience, as more or less "convenient." Thus, for instance, sounding a lot like Claude Lévi-Strauss, Lieberman argues that "evolution is a tinkerer, adapting existing structures that enhance reproductive success in the ever-changing conditions of life" (2000, 166). In arguing for the continuity of reason with our bodily life, Damasio also describes evolution as "thrifty and tinkering" (1994, 190).

21. Sacks calls such mimicking "pseudo-actions" and "simulacra of action and meaning": "Mrs. Y.'s tics," he writes in *Awakenings,* "*look* like actions or deeds—and not mere jerks or spasms or movements. One sees, for example, gasps, pants, sniffs, finger-snappings, throat-clearing, pinching movements, scratching movements, touching movements, etc., etc., which could all be part of a normal gestural repertoire. . . . These pseudo-actions, sometimes comic, sometimes grotesque, convey a deeply paradoxical feeling, in that they *seem* at first to have a definite (if mysterious) organization and purpose and then one realizes that in fact they do not" (1999, 109). Here Sacks is describing the kind of "pseudo-intentionality" that Lionel Essog describes in opposing the "accidental lunacy" of kissing to the seeming intention of his verbal tics. And this phenomena gets even more complicated in that those with Tourette syndrome often "camouflage," as my

scholarly friend told me, unintentional action with seeming intentional gestures. Lethem narrates such camouflage throughout *Motherless Brooklyn,* and Sacks describes it explicitly: "When I questioned Miss H. about this symptom [a lightening-quick movement of the right hand to the face] . . . she replied that it was 'a nonsense-movement.' . . . Within three days of its appearance, however, this tic had become associated with an intention and a use: it had become a mannerism, and was now used by Miss H. to adjust the position of her spectacles" (136).

4. The History of the Hand

1. Here is Jakobson's passage in its entirety: "The political slogan 'I like Ike /ay layk ayk/, succinctly structured, consists of three monosyllables and counts three diphthongs /ay/, each of them symmetrically followed by one consonantal phoneme, / . . l . . k . . k/. The makeup of the three words presents a variation: no consonantal phonemes in the first word, two around the diphthong in the second, and one final consonant in the third. . . . Both cola of the trisyllabic formula 'I like / Ike' rhyme with each other, and the second of the two rhyming words is fully included in the first one (echo rhyme), /layk/—/ayk/, a paronomastic image of a feeling which totally envelops its object. Both cola alliterate with each other, and the first of the two alliterating words is included in the second: /ay/—/ayk/, a paronomastic image of the loving subject enveloped by the beloved object. The secondary, poetic function of this electional catch phrase reinforces its impressiveness and efficacy" (1987, 70).

2. Merlin Donald offers a more elaborate definition (which still does not contradict Damasio's): consciousness, he writes, "is much deeper than the sensory stream. It is about building and sustaining mental models of reality, constructing meaning, and exerting autonomous intermediate-term control over one's thought process, even without the extra clarity afforded by having the explicit consensual systems of language. . . . Above all, this deliberative capacity is an active and causative element in cognition and a regulator and arbitrator of action" (2001, 75–76). Gerald Edelman studies consciousness by focusing more fully on neurology. After suggesting that consciousness is "what you lose when you fall into a deep dreamless sleep and what you regain when you wake up" (2005, 4), Edelman argues that "consciousness is a process, not a thing" and that "the process of consciousness is a dynamic accomplishment of the distributed activities of populations of neurons in many different areas of the brain" (6–7). These definitions are much more complex than simple sensory awareness, what Edelman calls "primary consciousness" (8).

3. Edelman makes the same distinction under the categories of primary consciousness and higher-order consciousness. "Primary consciousness," he writes, "is the state of being mentally aware of things in the world, of

having mental images in the present. It is possessed not only by humans but also by animals lacking semantic or linguistic capabilities whose brain organization is nevertheless similar to ours. Primary consciousness is not accompanied by any sense of a socially defined self with a concept of a past or a future. It exists primarily in the remembered present. In contrast, higher-order consciousness involves the ability to be conscious of being conscious, and it allows the recognition by a thinking subject of his or her own acts and affections. . . . At a minimal level, it requires semantic ability, that is, the assignment of meaning to a symbol. In its most developed form, it requires linguistic ability, that is, the mastery of a whole system of symbols and a grammar" (2005, 8–9). I suspect it is "core" or "primary" consciousness—Peirce's "icon"—that David Chalmers means by "experience."

4. For a systematic study of the relation between gesture and meaning—especially in childhood cognitive and linguistic development—see Goldin-Meadow 2003. Goldin-Meadow gives a wide-ranging survey of work focused on gestures in experimental psychology.

5. Of course, our evolutionary inheritance goes back to very ancient primates. "One of the best-known examples of a neural vestige in humans," Donald writes, "is the propriospinal tract, one of the spinal cord pathways that connects the brain to the hand. We have inherited this primitive hand-control pathway from a very ancient ancestor. It conveys messages from the motor brain to all the fingers of the hand at once, allowing only the diffuse contraction of all these fingers together. . . . Apes inherited this tract from distant monkey ancestors, but they (and we) are capable of much finer hand control because apes evolved a second spinal cord path for finer control, via the pyramidal tract. This newer pathway projects independently to each finger, enabling apes, and humans, to move one finger at a time and generate much finer patterns of coordinated hand movement. . . . In humans the propriospinal tract remains useful for a brief period during infancy, supporting a limited use of the hands for grabbing hold of things and people, until the pyramidal pathway matures" (2001, 106–7).

6. Frank Wilson describes the sense of "becoming one" with objects in the world: "The idea of 'becoming one' with a backhoe is no more exotic than the idea of a rider becoming one with a horse or a carpenter becoming one with a hammer, and this phenomenon itself may take its origin from countless monkeys who spent countless eons becoming one with tree branches. The mystical feel comes from the combination of a good mechanical marriage and something in the nervous system that can make an object external to the body feel as if it had sprouted from the hand, foot, or (rarely) some other place on the body where your skin makes contact with it" (1999, 63).

7. In logic, *intension* means the internal content of a notion or a concept as opposed to what it might refer to. Robert Scholes describes this term in relation to Peircean semiotics, specifically in contrasting Peirce and Saussure. "Peirce's 'tri-relative' notion of semiosis," he writes, "places him close to

Frege, to Carnap, and to Ogden and Richards and far from Saussure and his followers. We can display their terminologies in the following way, placing the comparable (but by no means identical) terms of each formulation in the same column" (1985, 92):

Frege	Expression *(Ausdruck)*	Sense *(Sinn)*	Reference *(Bedeutung)*
Carnap	Expression	Intension	Extension
Ogden/Richards	Symbol	Thought	Referent
Peirce	Sign	Interpretant	Object
Saussure	Signifier	Signified	—

8. Johnson suggests a physiological basis for this phenomenon. Some researchers, he notes, have isolated "mirror neurons" in chimps that fire when a chimp performs a particular activity (e.g., putting food in its mouth) *and also fire,* he writes, "when the monkey observed another monkey performing the task." Such "synchronic" firings for self and others, he speculates, might well be "the neurological root" of empathy, "which would mean that our skills were more than just an offshoot of general intelligence, but relied instead on our brains' being wired a specific way," and he further suggests that people suffering from autism might well "suffer from a specific neurological disorder that inhibits their ability to build theories of other minds" (2001, 196–202).

9. Toni Morrison narrates a wonderful example of the shared attention of narrative in *Beloved:* "Denver was seeing it now and feeling it—through Beloved. Feeling how it must have felt to her mother. Seeing how it must have looked. And the more fine points she made, the more detail she provided, the more Beloved liked it. So she anticipated the questions by giving blood to the scraps her mother and grandmother had told her—and a heartbeat. The monologue became, in fact, a duet as they lay down together, Denver nursing Beloved's interest like a lover whose pleasure was to overfeed the loved. . . . Denver spoke, Beloved listened, and the two did the best they could to create what really happened, how it really was" (1988, 78).

10. For a wonderful description of the phonic and linguistic basis of the bonding between parent and child, see Mithen 2006, chapter 6. Mithen describes the ways that "vocal gestures," closer to song than to language, situate and orient the child in the world. As in the automatic dysfunction of Tourette syndrome, functioning infants "demonstrate an interest in, and sensitivity to, the rhythms, tempos and melodies of speech long before they are able to understand the meanings of words" (69). Mithen also describes the automatic activity of laughter in relation to this bonding and notes that laughter—like Tourette syndrome—originates "in the evolutionarily oldest parts of the brain" (82). See also Provine 2000.

11. John Napier notes that evidence suggests "a right-handed bias in the Lower Paleolithic at least two hundred thousand years ago." Still, he observes that "the final surge that turned a right-handed bias into the right-handed dominance of modern humans is quite unknown" (1993, 128). Wilson presents a number of the theories concerning the adaptiveness of handedness (1999, 147–53). Trimble (2007) traces the relationship between the bicameral brain and poetry (especially religious poetry).

12. Gerald Edelman has created a model of "neural Darwinism" in which Darwin's notion of "population thinking"—which describes the way "functioning structures and whole organisms emerge as a result of selection among the diverse variant individual of a population, which compete with one another for survival" (2005, 33)—governs the development of individual brains by means of "neuronal group selection" among the vast "population" of neurons and neuronal pathways (33). This model might seem simply "internal," yet the neuronal group selection he describes responds to stimuli *outside* the brain (if not outside the body) to account for "the extraordinary amount of variation in each individual brain" (34).

13. Mithen's (2006) global argument is that "gestures, facial expressions, or tone, grunts, etc." were the functional elements of a protolanguage, based on its musical qualities, that served hominids for more than 200,000 years before the development of compositional language in *homo sapiens.* This early language, he argues, is closely related to music. Similarly, Donald (1991) argues that such a nongrammatical communicative system created the first stage of human consciousness in a "mimetic culture." See also Sacks 1989, 120.

14. Donald's tripartite division of consciousness into three levels (selective binding, short-term control, and intermediate and long-term governance) corresponds to his tripartite division of episodic perception, mimesis, and language he examined in his earlier book, *Origins of the Modern Mind,* which I discuss later. In that book, he notes that "an animal lacking a capacity for episodic memory and restricted to the procedural level would be not much more than a stimulus–response organism, a high-level automaton of the sort favored by the early behaviorists" (1991, 151). Descartes's mechanical doll, described in chapter 2, is such a "high-level automaton."

15. Throughout *The Right Mind,* Ornstein offers descriptions by patients with right-hemisphere damage of a painting by Norman Rockwell that depicts three people anxiously waiting in a physician's waiting room. Patients with right-hemisphere damage describe the painting as people at a baseball game ("they all seem so interested" [1997, 11]), or people at an unexciting Boy Scout meeting (26), or a young man "calling on his girl's parents" (43), or veterans in a church pew (63), or brothers watching television (80), or people at the movies (87). One patient simply lists the elements of the painting—position, clothing, faces—without suggesting any "larger" meaning (97). "If you were to look at a scene like the one in our

Norman Rockwell painting of the doctor's office," Ornstein concludes, "and you couldn't tell what it was, what the purpose of the room was, or why the people were waiting, then you might have difficulty deciding how to act, what to do. . . . And what might you think of someone walking into the room with a scalpel? You might respond as you would when attacked" (117). What is striking here is that observers with brain damage almost all fail to interpret the image *narratively* in terms of "why the people were waiting," the "aboutness" of a story I mention in chapter 1.

16. As I noted earlier, "sound-image" is Wade Baskin's translation of Saussure (1959); Harris calls it "sound pattern" (1983). Saussure goes to some pains to distinguish sounds from sound-images (or patterns), and he does so, I suspect, because he wants to emphasize what Peirce calls the indexical nature of the second articulation of language rather than the phenomenological experience of sound sensations, which Peirce describes under the category of icon. The apprehension of tone of voice as *indicative* of meaning—the right brain phenomenon Ornstein describes—is the apprehension of the *indexical* significance of *iconic* experience. If this is so, then the received opinion of Saussure's fully excluding the referential aspect of language (e.g., see note 7 with Robert Scholes's description of this situation) might need to be at least somewhat modified.

17. The sign systems developed in communities of deaf people are fully—*doubly*—articulated languages. For a fine discussion of such sign systems, see Sacks 1989. Frank Wilson cites the psycholinguist Harlan Lane describing Sign as linguistic systems that embody double articulation (though he does not use that term) (1999, 185–86).

18. Its simplicity consists in the symmetry between presence and absence, so that the presence or the absence of the phoneme /d/ in relation to "roe" and "road" have equal effect: presence and absence are not different *in kind*. (This is also true of spatial location in relation to "here" and "there.") The symbolic level is complex—it is, in Derrida's term, "irreducibly non-simple"—because the "addition" of information does not quite "delete" what was previously there, but it does not leave it unchanged either. Derrida makes much of this in relation to the linguistic category of "neutralization" I mention in chapter 2 (note 15). For a more elaborate discussion of Derrida's relationship to structural linguistics, including Martinet, see *Analogical Thinking* (2000a, esp. ch. 2).

19. Mithen's argument (2006)—which, in fact, favorably cites Corballis—is that language evolved from gestures that include "vocal gestures." Mithen emphasizes the *motor* nature of sound production.

20. Mithen offers a strong anthropological and historical argument for the function of "displacement"—which he defines as "reference to either the past or the future"—in relation to Donald's discussion of mimetic culture (2006, 316–17). Such "displacement," he says, "is a key feature of modern human language" (317). It is concerned with the future, although, unlike

Peirce's symbol, it does not "govern" the future in a lawlike way. In chapter 5, I examine Greimas's notion of "shifting out," which is his term for displacement.

21. By "sign" Wilson is referring to the sign languages developed in communities of deaf people, and he cites Sacks's study of sign communities, *Seeing Voices*. In that book, in a passage Wilson does not cite but that comes close to the phenomenon he is describing here, Sacks notes "that signing is not just the manipulation of symbols according to grammatical rules, but, irreducibly, the voice of the signer—a voice given special force, because it utters itself, so immediately, with the body. One can have or imagine disembodied speech, but one cannot have disembodied Sign" (1989, 119). Wilson cites the psycholinguist Harlan Lane who notes that "sign includes movements beyond what is doing in the hands—body shifts and limb movements and facial expressions—but still the core of it is in the hands" (1999, 185). In music—piano music comes particularly to mind, but our hands make music with oboes and bassoons as well—voice is "put out through the hand."

22. For this reason, Gould argues that the "units" or "individuals" of natural selection vary by levels. He names six "for convenience": "gene, cell lineage, organism, deme, species and clad" (2002, 73)—it is convenient because he can imagine other organizing principles. (In this he seems to agree with Dupré's pluralism, described in chapter 1, note 3.) I have been arguing for three levels, and Gould's catalog could be "conveniently" assimilated to my argument in terms of gene, organism, and species (i.e., Dawkins's "selfish gene," Darwin's "individual," and Gould's "punctuated equilibrium"). "Selection among cell-lineages," Gould writes, "although ancestrally important in the evolution of multicellular organisms, has largely been suppressed by the organismal level in the interests of its own integrity" (2002, 74); and *deme* is a particular interbreeding population within a species (later I note that scholars who argue for the natural selection of religion focus on the deme when they describe religious groups: see chapter 5, note 23). Finally, *clad* means "branch," and those who use this term contend that the parent species must be renamed, even if it does not change, when the daughter species branches off so that, Gould says, "many biologists reject (and regard as nonsense) the cladistic principle" (605). In these terms, cell-lineage might be (has been?) assimilated to individual organism, and deme (populations of organisms) might be assimilated to species (which encompasses or obviates clad), and my three levels—mechanistic genes, interactive individuals, and species-networks—are "convenient" indeed. In Gould's catalog, these "extra" categories seem to emphasize the particular *historical* activity of the classed phenomena of genes, organisms, and species.

23. Greimas's semiotic square, examined in chapter 2, describes presence and absence as different classes of fact. In *Rhetoric and Death* (1990), I elaborate at length the ways in which conceptions of life and death are not symmetrical.

24. Such a superimposition would align Lacan's "imaginary" with Peirce's index even though the "imaginary" seems more akin to Peirce's icon. What this suggests, I think, is that Lacan's imaginary takes up an icon for the purposes of *specifying;* it takes up icon as index. In any case, this *kind* of confusion underlines the creation of disorder out of order, which, to some degree, is the work of the symbolic and semiotic altogether, the conferring of randomness that McManus describes that, "at least in small amounts, can benefit complex systems" (2002, 229). In this superimposition, Lacan's "real" would align with Peirce's icon, unspeakable *sensational* phenomena.

5. Pain, Memory, and Religious Suffering

1. A proprioceptor, as the *Oxford English Dictionary* notes, is "any sensory structure which receives stimuli arising within the tissues (other, usually, than the viscera); esp. one concerned with the sense of position and movement of a part of the body." It is distinguished from an exteroceptor, "a sense organ which receives external stimuli, as those of touch," and an interoceptor, "any sensory receptor which receives stimuli arising within the body, or spec. within the viscera." Interoception, as Wall notes, is solely the province of C fiber nerve cells. Proprioception is the felt sense of the body in the world—it perceives, for instance, the different weight of two blocks held by each hand and thus measures comparative pressure.

The catalog of interoception, exteroception, and proprioception parallels Peirce's catalog of icon, index, and symbol. (It might well be that the similarity between proprioception and the symbol led Saussure to eliminate reference from his discussion of the sign.) The catalog also traces the three levels of scientific understanding I examine in chapter 2 and in these final three chapters focusing on the affect of poetry, the worldly work of narrative, and the subject of poetry.

2. Wall is citing a definition developed by a committee headed by Dr. Harold Merskey under the auspices of the International Association for the Study of Pain. The definition adds that in addition to being "associated with actual or potential tissue damage," pain may be "described in terms of such damage." "The IASP," Marni Jackson notes, "is now working on an addendum to avoid the possible suggestion that patients who can't articulate their pain—babies, or people with dementia, for instances—don't suffer pain" (2002, 21). This issue of the articulation of pain raises the problem of subjectivity in a manner related to the problem raised by Tourette syndrome where there is "articulation" but no intentional meaning.

3. Semiotics teaches us that even the "nothing" Saussure describes here can be a *material* phenomenon, analogous to the number zero I discuss in chapter 2 and the whole "negative science" it synecdochically represents.

4. Wall and Melzack developed gate control theory in the 1960s. "The

theory," Glucklich notes, "subsequently supported by numerous clinical studies, posits a mechanism in the spinal cord that controls the flow of neuronal stimuli from the body's peripheries to the brain, where pain registers. The gate-control mechanism operates by means of inhibiting signals that 'descend' from the brain to the gate and block the incoming signals" (2001, 52–53). Glucklich cites Melzack 1995. See also Melzack 1993; Melzack and Wall 1983, 222–39.

5. In his highly reductionist studies of memory focusing on single nerve cells (in the giant invertebrate sea snail aplysia), Eric Kandel examines the mechanisms of memory in terms of classical conditioning and "two nonassociative forms of learning: habituation and sensitization. . . . In habituation the animal learns to ignore a stimulus because it is trivial, whereas in sensitization it learns to attend to a stimulus because it is important." In classical conditioning, "learning occurs through the association of two stimuli or a stimulus and a response" (2006, 41).

6. Patrick Wall may well be describing the "objectlessness" of pain in its "uncomfortable puzzle" where "stimulus and response seem so strangely connected" (1999, 78). That connection, as I suggest later, might well be the manner in which the nociception of sensory nerves creates the phenomenal "illusion" of interoception. In any case, stimulus and response encompass the physiology and phenomenology I am examining.

7. The term itself was coined by Silas Weir Mitchell in his work with Civil War patients in America. Later he focused on diseases of women and is "revered in medical schools, where he is regarded as the father of modern neurology" (Morris 1993, 105). His "rest cure," which he prescribed for Charlotte Perkins Gilman, led to the composition of "The Yellow Wall Paper." For an extended discussion of Mitchell, see Jackson 2002, 193–200.

8. V. S. Ramachandran and Sandra Blakeslee note that Lord Nelson, with the greater authority of his own lost right arm, experienced phantom limb pains that led him "to proclaim that his phantom was 'direct evidence for the existence of the soul.' For if an arm can exist after it is removed," they paraphrase Nelson's reasoning, "why can't the whole person survive physical annihilation of the body?" (1998, 22–23; see also Jackson 2002, 72). They also describe the famous Victorian physicist Sir David Brewster, arguing for the existence of God from the phenomenon of "filling-in" the blind spot in vision (1998, 273).

9. Vertosick also argues that pain rather than pleasure is the "default" mode: "Pleasure is a much more complicated cognitive process relative to pain. Bogus signals emanating from unemployed spinal neurons don't have the complexity to register in our minds as pleasant" (2000, 48).

10. Ramachandran and Blakeslee describe a kind of "pain memory" associated with phantom pain. "Some patients," they write, "say that the pain they felt in their limbs immediately prior to amputation persists as a kind of pain memory. For example, soldiers who have grenades blow up in their

hands often report that their phantom hand is in a fixed position, clenching the grenade, ready to toss it. The pain in the hand is excruciating—the same they felt the instant the grenade exploded, seared permanently in their brains" (1998, 51). See also Melzack for a discussion of pain memories (1993, 627).

11. For a sense of the working of narrative and discourse in relation to phenomenology, see Holenstein (1976). Throughout the preceding chapters, I have cited Stephen Jay Gould's discussion of history, which is episodic memory writ large. As part of his argument, he offers a striking examination of the difference between the "predictability" of mechanics and the "contingency" of historical explanation in his discussion of the study of history—including evolution—as explanation "*after* the fact" (1989, 51). He argues that he is "not speaking of randomness . . . , but of the central principle of all history—*contingency.* A historical explanation does not rest on direct deductions from laws of nature, but on an unpredictable sequence of antecedent states, where any major change in any step of the sequence would have altered the final result. The final result is therefore dependent, or contingent, upon everything that came before—the unerasable and determining signature of history" (283). As we saw in chapter 2, he rests this antireductionist argument on a conception of Whewell's notion of consilience (282), which is very different from that of E. O. Wilson's.

12. In a different schema, Aristotle, as Morris notes, categorized pain as one manifestation of emotion. Noting that pain impulses are directed to the limbic system as well as the cerebral cortex via the thalamus, Morris suggests that "Aristotle thus was not so far off in classifying pain as an emotion" (1993, 155). Damasio argues that pain, in fact, "does not qualify for emotion," though "the sensation of pain . . . may also induce emotions on its own" (1999, 71). It is my contention in the larger argument of *Intangible Materialism* that pain can be understood in reductionist mechanistic terms, such as those Kandel has so successfully pursued, in terms of the contingencies of evolutionary history, and in relation to the contrary-to-fact discourse of semiotics. In this catalog, basic emotions—fear, joy, surprise, disgust, anger—take their place in the context of evolutionary adaptations.

13. Physicians are trained to make this abstraction of pain from its cognitive and emotional meaning as well, and in Chalmers's argument, so do reductive "physicalist" accounts. For a discussion of the opposition between logico-scientific knowledge and narrative knowledge, see Vannatta, Schleifer, and Crow 2005.

14. Earlier, Atran argues that "metarepresentation . . . involves the ability to track and build a notion of self over time, to model other minds and worlds, and to represent beliefs about the actual world as being true or false. It also makes lying and deception possible" (2003, 15). In this definition, Atran comes close to Umberto Eco's defining description of semiotic systems in general: "Semiotics," Eco writes, "is concerned with everything

that can be *taken* as a sign. A sign is everything which can be taken as significantly substituting for something else. . . . Thus *semiotics is in principle the discipline studying everything which can be used in order to lie.* If something cannot be used to tell a lie, conversely it cannot be used to tell the truth: it cannot in fact be used 'to tell' at all" (1976, 7). In his book *Why We Lie: The Evolutionary Roots of Deception and the Unconscious Mind,* David Livingston Smith argues that "one of the basic functions of language is to deceive others" (2004, 7), and he further argues that "the power to deceive is our main weapon in the struggle for social survival" (150).

15. Dennett offers two definitions of religion, by Emil Durkheim and Clifford Geertz, that do not include supernatural agents (2006, 391; see also p. 198, for "'godless' religions"). Sam Harris makes a similar argument, offering the example of Buddhism as nonsupernatural religion (2005, ch. 7).

Throughout his study, Atran also suggests the "unnaturalness" of religion under the categories of counterfactual and counterintuitive: "Everybody, whether they are religious or not, implicitly knows that religion is costly, counterfactual, and even counterintuitive" (2002, 5). Atran's term *counterfactual* underlines the connection between religious experience (which he is studying) and the experience of literature. His second term, *counterintuitive,* might even call our attention to the delight literature affords us.

16. More generally, Norman Cook argues that "at every level of linguistic processing that has been investigated experimentally, the right hemisphere has been found to make characteristic contributions, from the processing of affective aspects of intonation, through the appreciation of word connotations, the decoding of the meaning of metaphors and figures of speech, to the understanding of the overall coherency of verbal humour, paragraphs and short stories" (2002, 169).

17. Earlier, Ornstein offers a materialist narrative depicting what "had to spur the pre-human brain on to greater size. That spur," he writes, "might have been packaging requirements given the need to keep the delicate internal brain cells from overheating in the African sun [in the now-upright hominids]" (1997, 32). Increased blood supply, necessary for cooling the brain, also "enabled brain volume to increase"; moreover, the increased numbers of neurons were not only more numerous "but more insulated than our forebears'. Because of the extra insulation, we can make connections more easily. . . . These extra cells and connections, once in place, could then continue the basic plan of asymmetrical [hemispheric brain] development that originated in earlier-evolved animals. There were more cells available, and the brain, freed of having to make the forelimbs similar, could exercise those cells differently" (33).

18. We note in *Culture and Cognition* that a psychoanalyst like Jacques Lacan "and Continental semiotics more generally, too quickly dispense

with . . . referential frameworks, taking the special case of fiction-writing [shifting out] to be the defining case of cognition" (Schleifer, Davis, and Mergler 1992, 143). An important aspect of Peircean semiotics, as I suggested in the previous chapter, is its inclusion of the indexical-referential function. Valentine Daniel makes a similar point in his discussion of his use of Peirce in anthropological studies (1987, 14–19) in the fascinating analysis I cite later in the chapter. For a remarkably detailed account of the coevolution of language and the brain, see Deacon 1998.

19. The possibility of two different kinds of attention they bring to experience is also at the heart of the systematic disorder of semiosis, with its hierarchy of units of communication and the negation of that hierarchy Greimas observes (1983, 82).

20. This should not suggest that pain and suffering can be understood in terms of the opposition between physical and mental pain. As Morris argues, the "absolute dualism of mental and physical pain is a comparatively recent idea whose time, as I contend, has long passed. A truly effective dialogue between medicine and literature may just succeed in driving a stake through its heart" (1993, 27). Morris goes on to say that "human pain similarly lives within us only as we have brains and minds to perceive it, only as we possess intellect and emotions to register its depth, source, and implications. In perceiving our pain, we transform it from a simple sensation into the complex mental-emotional events that psychologists and philosophers call perception" (29; see also Morris 1998, 118). In a similar fashion, Glucklich argues that "it is impossible to separate the confluence of psychological, social, and cultural factors from the physical event in the overall experience of the pain" (2001, 52). He bases his argument on the gate control theory propounded by Melzack and Wall, which "has led to the recognition that the intensity, duration, and nature of pain depend on decisions of the central brain not just peripheral stimulation" (53). Cassell's catalog includes elements that are iconic and physiological ("existence below awareness," "a body"), indexical and bioadaptive ("family," "culture and society," "associations with others," politics), and symbolic and semiotic ("personality," "character," "roles," politics [again], "a secret life, a believed-in future, and a transcendent dimension").

21. In a subsequent book, Dunbar examines the limit of orders of intentionality, concluding that understanding beyond seventh-order intentionality is virtually impossible and that in experiments, "around fifth order" there is a "sudden collapse" of comprehension (2004, 45–52). He also reviews, skeptically, the "controversial" evidence of other species possessing theory of mind (55–62).

22. In a chapter focused on the logic of narrative, in *Culture and Cognition,* my colleagues and I focus on the special relationship between sender and receiver in Greimassian semiotics—two of six narrative roles (or "actants") he derives from Propp and from grammatical categories—and we

argue that the narrative actor who also assumes the role of receiver defines the particular genre of a literary work by obtaining the wished-for good of narrative (Schleifer, Davis, and Mergler 1992, 64–95). Tragedy is marked by the fact that the helper of the hero is also the receiver; melodrama is marked by the fact that the hero is also the receiver, living, so to speak, to tell his or her own tale. In this conception, particular genres embody the work of narrative, the point of a story. The sender, as I mentioned in chapter 1, can always be universalized—in religion, this often is transcendental god-agency—but the receiver of the narrative's value underlines the worldly work of narrative, its point, its agency in the world.

23. Besides her visionary experience, Teresa was an effective and important organizer of Spanish institutions for nuns. David Sloan Wilson, in his study of religion and evolution, emphasizes the degree to which religion is a social institution as well as a site for personal experience. Religion, for Wilson, is "communal" as well as "otherworldly" (2003, 41), and he argues that religious thought and evolutionary thought "both spring from the fundamental problem of social life and its partial solution that lies at the heart of religion and which can be explained by multilevel selection theory" (39). By "multilevel selection theory," he means the ability of groups to function as adaptive units. (See chapter 4, note 22, for a discussion of Gould's multilevel selection theory. The group Wilson describes is an example of Gould's level of the deme.) Later, Wilson states that "religions exist primarily for people to achieve together what they cannot achieve alone. The mechanisms that enable religious groups to function as adaptive units include the very beliefs and practices that make religion appear enigmatic to so many people who stand outside of them" (159–60). The "communal" and "fundamental problem of social life" are modes of being, as I suggested in chapter 4, related to Peirce's index. Boyer argues, in a vein similar to Wilson's, that one should study not the religious experience of "exceptional people" (e.g., "mystics and visionaries") as William James and many others have done (2001, 309), but the less exceptional and dramatic religious experiences of ordinary people in order to discover the "common potential" from which exceptional experiences develop (310). The dialectic I explore here between Teresa's visionary experience and Stafford's allusiveness in her story about the extraordinary but nonsupernatural pain of her character might help discover the "common potential" of transforming pain to suffering.

My friend and colleague Tom Boyd once announced in a class we were teaching together, Rhetoric and Religion, that he was a Christian but did not believe that the work of the religion was first and foremost the salvation of the individual soul; rather, it was community building. We were all amazed.

24. Throughout more scientific descriptions of religious experience—both neurological and evolutionary descriptions—the phenomenon of out-of-body experience is repeatedly described. See, for instance, d'Aquili and

Newberg (1999) and Atran (2002, 190). Sacks begins *Musicophilia* (2007) with a case history of an out-of body experience.

25. As I already mentioned, Descartes also claimed that the phenomenon of phantom pain demonstrated the existence of another version of personhood, "soul."

26. Glucklich qualifies this assertion by emphasizing the phenomenal nature of the "self": even though "the body-image is a phantom, we do in fact experience ourselves *as though* possessing an integrated and concrete body-self. The essence of phenomenology is that 'as though' experiences—phantoms—can acquire a reality that seems more vivid and concrete than what is 'truly so'" (2001, 90). He also says, "Very much like the body in the mirror [in Lacan's 'mirror stage'], the body-image is itself an abstraction: In reality it is never a thing—not even a constant phenomenon—let alone an entity. It shifts endlessly with the changes and movements of the body" (90).

27. As I note in chapter 1, Peirce himself describes these three categories more generally. Let me repeat Peirce's description. "My view," he writes, "is that there are three modes of being, and [I] hold that we can directly observe them in elements of whatever is at any time before the mind in any way. They are the being of positive qualitative possibility, the being of actual fact, and the being of law that will govern the future" (1931–35, 1:23; Daniel cites this passage as well [1987, 239]). This is why, for Peirce, "the modal nature of Thirdness is that future conditional that must be regarded as a real potentiality both in our experience and in nature itself" (Daniel 1987, 244). The future conditional is a subset of contrary-to-fact statements.

28. This process almost completely reverses David Chalmers's description of the consciousness mind (1996).

29. Firstness, Peirce asserts, "is present and immediate" and thus "avoids being the object of some sensation. It precedes all synthesis and all determination. It has no unity and no parts" (1931–35, 1:357).

30. The banishment of reference—of Peirce's index—also banishes time and the materiality of evolution from Saussurean linguistics, which requires an almost apocalyptic advent of meaning into the world. Still, as I note in chapter 4 (see note 16), this judgment of Saussure might well be modified.

31. Hillis Miller describes—and to some degree historicizes—the identification of literature with the sacred in "The Search for Grounds in Literary Studies" (1985). He identifies four "grounds" for literary study, linguistic, psychological, historical, and—though he does not quite use the words—religious or *anagogic*. I prefer "anagogic" because it suggests the overwhelming experience of meaning.

32. Here I should repeat what I cited in the Introduction, that in his accounts of quantum physics, Heisenberg replaces notions of simultaneity with the Copenhagen interpretation of alternation: Complementarity encourages physicists "to apply alternatively different classical concepts

which would lead to contradictions if used simultaneously" (Heisenberg 1958, 179).

33. Later, Daniel notes that "whereas Firstness designates that category of pure quality or even a pure qualitative possibility considered in abstraction from everything else, and whereas Secondness represents sheer existence, brute fact, or actuality, Thirdness is that gentle force that mediates First and Second, bringing them into significant relationship. *Habit* represents Thirdness almost to perfection" (1987, 27).

34. Weil, unlike Pansy, understands that such affliction can be understood in relation to Cassell's "transcendent dimension," that it is, in relation to Catholic theological thinking, all that one can bring before God at Judgment. Flannery O'Connor understands this as well. In her story, "The Artificial Nigger," the protagonist, Mr. Head, "stood appalled, judging himself with the thoroughness of God, while the action of mercy covered his pride like a flame and consumed it. He had never thought himself a great sinner before but he saw now that his true depravity had been hidden from him lest it cause him despair. He realized that he was forgiven for sins from the beginning of time, when he had conceived in his own heart the sin of Adam, until the present when he had denied poor Nelson. He saw that no sin was too monstrous for him to claim as his own, and since God loved in proportion as He forgave, he felt ready at that instant to enter Paradise" (1972, 269–70). For a discussion of this experience in O'Connor, see Schleifer 1993.

35. Aristotle's literary use of this medical and physiological term *catharsis* is explored in *Medicine and Humanistic Understanding*, where three possible translations—purgation, purification, and clarification—are aligned with three definitions of health (see Vannatta, Schleifer, and Crow 2005, IV, hypertext to p. 29). "All three definitions of 'catharsis,'" we note, "are closely related to medicine: its practices of healing; its objective, scientific understandings; and its global enterprise of confronting suffering in the person of its human sufferer with pity and empathy and confronting suffering in its various causes with science and care. These definitions also correspond to various definitions of health we have encountered: (#1) health as the absence of disease; (#2) health as what the WHO calls 'the total well-being of body and mind'; and (#3) John Stone's definition of health in his poem, 'House Call,' as 'whatever works / and for as long.'" These three definitions also fit within the three levels of scientific understanding I describe in this book.

Bibliography

Aggleton, John, and Andrew Young. 2000. "The Enigma of the Amygdala: On Its Contribution to Human Emotion." In *Cognitive Neuroscience of Emotion,* ed. Richard Lane and Lynn Nadel, 106–28. Oxford: Oxford University Press.

Andersen, Hanne, Peter Barker, and Xiang Chen. 2006. *The Cognitive Structure of Scientific Revolutions.* Cambridge: Cambridge University Press.

Arnold, Matthew. 1868. "The Function of Criticism at the Present Time." In *Lectures and Essays in Criticism,* ed. Robert Henry Super. Ann Arbor: University of Michigan Press, 1980.

Arrighi, Giovanni. 1994. *The Long Twentieth Century: Money, Power, and the Origins of Our Time.* London: Verso.

Atran, Scott. 2002. *In Gods We Trust: The Evolutionary Landscape of Religion.* New York: Oxford University Press.

Attridge, Derek. 1988. *Peculiar Language.* Ithaca, N.Y.: Cornell University Press.

Auden, W. H. 1966. *Collected Shorter Poems, 1927–1957.* New York: Random House.

Bakhtin, M. M. [V. N. Vološinov]. 1987. "Discourse in Life and Discourse in Art (Concerning Sociological Poetics)." In *Freudianism: A Critical Sketch,* trans. I. R. Titunik, 93–116. Bloomington: Indiana University Press.

Barreca, Regina. 1993. "Voodoo." In *Death and Representation,* ed. Sarah Goodwin and Elisabeth Bronfen. Baltimore: Johns Hopkins University Press.

Barthes, Roland. 1970. *Writing Degree Zero.* Trans. Annette Lavers and Colin Smith. Boston: Beacon Press.

Bataille, Georges. 1985. *Visions of Excess: Selected Writings, 1927–1939.*

Trans. Allan Stoekl, with Carl R. Vovitt and Donald M. Leslie Jr. Minneapolis: University of Minnesota Press.

Bell, John S. 1964. "On the Einstein-Poldolsky-Rosen Paradox," *Physics* 1:195.

Benjamin, Walter. 1996. "On Language as Such and on the Language of Man." Trans. Edmund Jephcott. In *Selected Writings: Volume 1, 1913–1926,* ed. Marcus Bullock and Michael W. Jennings, 62–74. Cambridge, Mass.: Harvard University Press.

Blitz, David. 1992. *Emergent Evolution: Qualitative Novelty and the Levels of Reality.* New York: Springer.

Bogan, James, and James Woodward. 1988. "Saving the Phenomena." *Philosophical Review* 97:303–52.

Bohm, David. 1986. Interview. In Davies and Brown 1986a, 118–34.

Bowen, Elizabeth. 1962. *The Heat of the Day.* New York: Penguin Books.

Bowker, John. 1970. *Problems of Suffering in Religions of the World.* Cambridge: Cambridge University Press.

Boyer, Pascal. 2001. *Religion Explained: The Evolutionary Origins of Religious Thought.* New York: Basic Books.

Brent, Joseph. 1998. *Charles Sanders Peirce: A Life.* Bloomington: Indiana University Press.

Bridgman, Percy. 1941. *The Nature of Thermodynamics.* Cambridge, Mass.: Harvard University Press.

Bruun, Ruth. 1988. "The Natural History of Tourette's Syndrome." In Cohen, Bruun, and Leckman 1988, 21–39.

Burke, Kenneth. 1994. "Literature as Equipment for Living." In *Contemporary Literary Criticism,* ed. Robert Con Davis and Ronald Schleifer, 3rd ed. New York: Longman.

Burrell, David. 1973. *Analogy and Philosophic Language.* New Haven, Conn.: Yale University Press.

Cameron, Sharon. 1979. *Lyric Time: Dickinson and the Limits of Genre.* Baltimore: Johns Hopkins University Press.

Campbell, Jeremy. 1984. *Grammatical Man: Information, Entropy, Language and Life.* Harmondsworth, England: Penguin.

Capra, Fritjof. 1997. *The Web of Life: A New Synthesis of Mind and Matter.* London: HarperCollins.

Cassell, Eric. 1991. *The Nature of Suffering and the Goals of Medicine.* New York: Oxford University Press.

Cassirer, Ernst. 1951. *The Philosophy of the Enlightenment.* Trans. Fritz Kolln and James Pettegrove. Princeton, N.J.: Princeton University Press.

Casti, John. 1995. *Complexification: Explaining a Paradoxical World through the Science of Surprise.* New York: Harper Perennial.

Chalmers, David. 1996. *The Conscious Mind: In Search of a Fundamental Theory.* Oxford: Oxford University Press.

Chandler, Daniel. 1994–2008. *Semiotics for Beginners*. Aberystwyth: University of Wales. http://www.aber.ac.uk/media/Documents/S4B.

Chapman, Richard. 1986. "Pain, Perception, and Illusion." In *The Psychology of Pain,* ed. Richard A. Sternbach, 2nd ed., 153–79. New York: Raven Press.

Charon, Rita, and Maura Spiegel. 2005. "On Conveying Pain/On Conferring Form." *Literature and Medicine* 24: vi–ix.

Clarke, Bruce. 2002. "From Thermodynamics to Virtuality." In *From Energy to Information: Representation in Science and Technology, Art, and Literature,* ed. Bruce Clarke and Linda Dalrymple Henderson, 17–34. Stanford, Calif.: Stanford University Press.

Cohen, Donald J., Ruth Bruun, and James Leckman, eds. 1988. *Tourette's Syndrome and Tic Disorders: Clinical Understanding and Treatment.* New York: Wiley.

Conrad, Joseph. 1924. Preface to *The Nigger of the "Narcissus."* New York: Doubleday, Page.

Cook, Norman. 2002. "Bihemispheric Language: How the Two Hemispheres Collaborate in the Processing of Language." In *The Speciation of Modern Homo Sapiens,* ed. T. J. Crow, 169–96. Oxford: Oxford University Press.

Corballis, Michael. 2002. *From Hand to Mouth: The Origins of Language.* Cambridge, Mass.: MIT Press.

Crane, Hart. 1966. *The Complete Poems and Selected Letters and Prose.* New York: Doubleday Anchor.

Crease, Robert, and Charles Mann. 1986. *The Second Creation: Makers of the Revolution in Twentieth-Century Physics.* New York: Macmillan.

Culler, Jonathan. 1976. *Saussure.* Glasgow: Fontana/Collins.

———. 1983. "Convention and Meaning: Derrida and Austin." *New Literary History* 13:15–30.

Damasio, Antonio. 1994. *Descartes' Error: Emotion, Reason, and the Human Brain.* New York: Avon.

———. 1999. *The Feeling of What Happens: Body and Emotion in the Making of Consciousness.* New York: Harcourt.

———. 2003. *Looking for Spinoza: Joy, Sorrow, and the Feeling Brain.* New York: Harcourt.

Daniel, E. Valentine. 1987. *Fluid Signs: Being a Person the Tamil Way.* Berkeley: University of California Press.

d'Aquili, Eugene, and Andrew Newberg. 1999. *The Mystical Mind: Probing the Biology of Religious Experience.* Minneapolis: Fortress Press.

Davies, Paul [P. C. W.], and John Gribbin. 1992. *The Matter Myth: Dramatic Discoveries That Challenge Our Understanding of Physical Reality.* New York: Touchstone Books.

Davies, P. C. W., and J. R. Brown, eds. 1986a. *The Ghost in the Atom.* Cambridge: Cambridge University Press.

———. 1986b. "The Strange World of the Quantum." In Davies and Brown 1986a, 1–39.

Davis, Robert Con, and Ronald Schleifer. 1991. *Criticism and Culture: The Role of Critique in Modern Literary Theory.* Harlow, England: Longman.

Dawkins, Richard. 2006. *The God Delusion.* Boston: Houghton Mifflin.

Deacon, Terrence. 1998. *The Symbolic Species: The Co-evolution of Language and the Brain.* New York: Norton.

Dear, Peter. 1995. *Discipline and Experience: The Mathematical Way in the Scientific Revolution.* Chicago: University of Chicago Press.

Deleuze, Gilles. 1983. *Nietzsche and Philosophy.* Trans. Hugh Tomlinson. New York: Columbia University Press.

de Man, Paul. 1979. *Allegories of Reading.* New Haven, Conn.: Yale University Press.

———. 1996. "Pascal's Allegory of Persuasion." In *Aesthetic Ideology,* ed. Andrzej Warminski, 51–69. Minneapolis: University of Minnesota Press.

Dennett, Daniel. 2006. *Breaking the Spell: Religion as a Natural Phenomenon.* New York: Viking.

Depew, David. 1994. "Review: Understanding Charles Sanders Peirce (In Some Ways)" [a review of *Charles Sanders Peirce: A Life* by Joseph Brent]. *Reviews in American History* 22:119–24.

Derrida, Jacques. 1982. *Margins of Philosophy.* Trans. Alan Bass. Chicago: University of Chicago Press.

———. 2001. "Typewriter Ribbon: Limited Ink (2) ('within such limits')." Trans. Peggy Kamuf. In *Material Events: Paul de Man and the Afterlife of Theory,* ed. Tom Cohen, Barbara Cohen, J. Hillis Miller, and Andrzej Warminski, 277–360. Minneapolis: University of Minnesota Press.

Descartes, René. 1985. *The Philosophical Writings,* vol. 1. Trans. John Cottingham, Robert Stoothoff, and Dugald Murdoch. New York: Cambridge University Press.

Diamond, Jared. 1992. *The Third Chimpanzee: The Evolution and Future of the Human Animal.* New York: Harper Perennial.

Donald, Merlin. 1991. *Origins of the Modern Mind: Three Stages in the Evolution of Culture and Cognition.* Cambridge, Mass.: Harvard University Press.

———. 2001. *A Mind So Rare: The Evolution of Human Consciousness.* New York: Norton.

Doyle, Roddy. 1996. *The Woman Who Walked into Doors.* New York: Viking.

Dunbar, Robin. 1996. *Grooming, Gossip, and the Evolution of Language.* Cambridge, Mass.: Harvard University Press.

———. 2004. *The Human Story: A New History of Mankind's Evolution.* London: Faber and Faber.
Dupré, John. 1993. *The Disorder of Things: Metaphysical Foundations of the Disunity of Science.* Cambridge, Mass.: Harvard University Press.
Eagleton, Terry. 1983. *Literary Theory: An Introduction.* Minneapolis: University of Minnesota Press.
Eco, Umberto. 1976. *A Theory of Semiotics.* Bloomington: Indiana University Press.
———. 1983. "Horns, Hooves, and Insteps: Some Hypotheses on Three Types of Abduction." In *The Sign of the Three,* ed. Umberto Eco and Thomas Sebeok. Bloomington: Indiana University Press.
Edelman, Gerald. 2005. *Wider Than the Sky: The Phenomenal Gift of Consciousness.* New Haven, Conn.: Yale University Press.
Einstein, Albert, Boris Podolsky, and Nathan Rosen. 1935. "Can Quantum-Mechanical Description of Physical Reality Be Considered Complete?" *Physics Review* 47:777.
Eliot, George. 1961. *The Mill on the Floss.* Boston: Houghton Mifflin.
Eliot, T. S. 1964. *Selected Poems.* New York: Harvest/HBJ Books.
———. 1974. "Tradition and the Individual Talent." In *Selected Prose,* ed. Frank Kermode, 37–44. New York: Harcourt Brace Jovanovich.
Emery, Nathan, and David Amaral. 2000. "The Role of the Amygdala in Primate Social Cognition." In *Cognitive Neuroscience of Emotion,* ed. Richard Lane and Lynn Nadel, 156–91. Oxford: Oxford University Press.
Ermarth, Elizabeth Deeds. 1983. *Realism and Consensus in the English Novel.* Princeton, N.J.: Princeton University Press.
Fahn, Stanley, and Gerald Erenberg. 1988. "Differential Diagnosis of Tic Phenomena: A Neurologic Perspective." In Cohen, Bruun, and Leckman 1988, 41–54.
Felman, Shoshana. 2002. *The Scandal of the Speaking Body: Don Juan with J. L. Austin, or Seduction in Two Languages.* Trans. Catherine Porter. Stanford, Calif.: Stanford University Press.
Fische, Max. 1981. Introductory note. In Thomas Sebeok, *The Play of Musement.* Bloomington: Indiana University Press.
Fischer, David Hackett. 1996. *The Great Wave: Price Revolutions and the Rhythm of History.* New York: Oxford University Press.
Flynn, Maureen. 1996. "The Spiritual Uses of Pain in Spanish Mysticism." *Journal of the American Academy of Religion* 64:257–78.
Foucault, Michel. 1972. *The Archaeology of Knowledge.* New York: Harper and Row.
Fowler, Alastair. 1982. *Kinds of Literature: An Introduction to the Theory of Genres.* Cambridge, Mass.: Harvard University Press.

Fraser, J. T. 1982. *The Genesis and Evolution of Time: A Critique of Interpretation in Physics.* Brighton, England: Harvester Press.
Freadman, Anne. 2004. *The Machinery of Talk: Charles Peirce and the Sign Hypothesis.* Stanford, Calif.: Stanford University Press.
Freud, Sigmund. 1997. *Dora: An Analysis of a Case of Hysteria.* New York: Touchstone Books.
Gaukroger, Stephen. 1995. *Descartes: An Intellectual Biography.* New York: Oxford University Press.
Geertz, Clifford. 1973. *The Interpretations of Cultures.* New York: Harper Torchbooks.
Glucklich, Ariel. 2001. *Sacred Pain: Hurting the Body for the Sake of the Soul.* New York: Oxford University Press.
Godzich, Wlad. 1986. "Foreword: The Tiger on the Paper Mat." In Paul de Man, *The Resistence to Theory,* ix–xviii. Minneapolis: University of Minnesota Press.
Goldin-Meadow, Susan. 2003. *Hearing Gesture: How Our Hands Help Us Think.* Cambridge, Mass: Harvard University Press.
Goodall, Jane. 1986. *The Chimpanzees of Gombe: Patterns of Behavior.* Cambridge, Mass.: Harvard University Press.
Gould, Stephen Jay. 1980. *The Panda's Thumb.* New York: Norton.
———. 1986. "Evolution and the Triumph of Homology, or Why History Matters." *American Scientist* 74 (January–February): 60–69.
———. 1989. *Wonderful Life: The Burgess Shale and the Nature of History.* New York: Norton.
———. 2002. *The Structure of Evolutionary Theory.* Cambridge, Mass.: Harvard University Press.
Gray, J. A. 1982. *The Neuropsychology of Anxiety: An Enquiry into the Functions of the Septo-Hippocampal System.* Oxford: Oxford University Press.
———. 1994. "Framework for a Taxonomy of Psychiatric Disorder." In *Emotions: Essays on Emotion Theory,* ed. Stephanie Van Gooze, Nanne Van de Poll, Joseph Sergeant, 29–59. Hillsdale, N.J.: Erlbaum.
Greenblatt, Stephen. 1995. "Culture." In *Critical Terms for Literary Study,* ed. Frank Lentricchia and Thomas McLaughlin, 225–32. Chicago: University of Chicago Press.
Greimas, A. J. 1963. "La linguistique statistique et la linguistique structural." *Le Français moderne* 31:55–68.
———. 1970. "La linguistique structurale et la poétique." In *Du Sens,* 271–83. Paris: Seuil.
———. 1972. Introduction to *Essais de Sémiotique Poétique,* ed. A. J. Greimas, 6–24. Paris: Librairie Larousse.
———. 1983. *Structural Semantics.* Trans. Danielle McDowell, Ronald Schleifer, and Alan Velie. Lincoln: University of Nebraska Press.
Greimas, Algirdas Julien, and Joseph Courtés. 1982. *Semiotics and Lan-*

guage: An Analytical Dictionary. Trans. Larry Crist, Daniel Patte, et al. Bloomington: Indiana University Press.

Harman, Gilbert. 1965. "The Inference to the Best Explanation." *Philosophical Review* 74: 88–95.

Harris, Sam. 2005. *The End of Faith*. New York: Norton.

Hatzfeld, Helmut. 1969. *Santa Teresa de Avila*. New York: Twayne.

Hawthorne, James, and Michael Silberstein. 1995. "For Whom the Bell Arguments Toll." *Synthese* 102: 99–138.

Hayles, N. Katherine. 1991. "Constrained Constructivism: Locating Scientific Inquiry in the Theater of Representation." *New Orleans Review* 18: 76–85.

———. 1999. *How We Became Posthuman: Virtual Bodies in Cybernetics, Literature, and Informatics*. Chicago: University of Chicago Press.

Heidegger, Martin. 1967. *What Is a Thing?* Trans. W. B. Barton and Vera Deutsch. Chicago: Henry Regnery Company.

Heisenberg, Werner. 1952. *Philosophical Problems of Nuclear Science*. Trans. F. C. Hayes. New York: Fawcett World Library.

———. 1958. *Physics and Philosophy*. New York: Harper Torchbooks.

Hilgard, Ernest. 1977. *Divided Consciousness: Multiple Controls in Human Thought and Action*. New York: Wiley.

Hilts, Philip. 1995. *Memory's Ghost: The Strange Tale of Mr. M. and the Nature of Memory*. New York: Simon and Schuster.

Hjelmslev, Louis. 1961. *Prolegomena to a Theory of Language*. Trans. Francis Whitfield. Madison: University of Wisconsin Press.

Hofstadter, Douglas. 1979. *Gödel, Escher, Bach: An Eternal Golden Braid*. New York: Basic Books.

Hogan, Patrick. 2003. *The Mind and Its Stories: Narrative Universals and Human Emotion*. Cambridge: Cambridge University Press.

Holenstein, Elmar. 1976. *Roman Jakobson's Approach to Language*. Trans. Catherine Schelbert and Tarcisius Schelbert. Bloomington: Indiana University Press.

Holland, John H. 1998. *Emergence: From Chaos to Order*. Reading, Mass.: Perseus Books.

"How You Feel Pain." 2007. Mayo Foundation for Medical Education and Research. http://mayoclinic.com/health/pain/PN00017.

Hunter, Kathryn. 1991. *Doctors' Stories*. Princeton, N.J.: Princeton University Press.

Jackson, Marni. 2002. *Pain: The Fifth Vital Sign*. New York: Crown Publishers.

Jakobson, Roman. 1987. "Linguistics and Poetics." In *Language in Literature*, 62–94. Cambridge, Mass.: Harvard University Press.

Johnson, Harold. 1973. "Changing Concepts of Matter from Antiquity to Newton." In *Dictionary of the History of Ideas: Studies of Selected Pivotal Ideas*, ed. Philip Wiener, 3:185–96. New York: Scribner.

Johnson, Steven. 2001. *Emergence: The Connected Lives of Ants, Brains, Cities, and Software*. New York: Touchstone Books.

Kandel, Eric. 2006. *In Search of Memory: The Emergence of a New Science of Mind*. New York: Norton.

Kaplan, Robert. 1999. *The Nothing That Is: A Natural History of Zero*. New York: Oxford University Press.

Keats, John. 1967. "This Living Hand." In *English Romantic Writers,* ed. David Perkins, 1205. New York: Harcourt, Brace, & World.

Keller, Helen. 1908. *The World I Live In*. New York: Century.

Kenner, Hugh. 1978. *Joyce's Voices*. Berkeley: University of California Press.

Kim, Jaegwon. 1993. "The Non-Reductivist's Troubles with Mental Causation." In *Mental Causation,* ed. John Heil and Alfred Mele, 189–210. Oxford: Oxford University Press.

Klee, Robert. 1984. "Micro-Determinism and Concepts of Emergence." *Philosophy of Science* 51:44–63.

Kushner, Howard I. 1999. *A Cursing Brain? The Histories of Tourette Syndrome*. Cambridge, Mass.: Harvard University Press.

Lacan, Jacques. 1972. "Seminar on 'The Purloined Letter.'" Trans. Jeffrey Mehlman. *Yale French Studies* 48:38–72.

Latour, Bruno. 1986. "Visualization and Cognition: Thinking with Eyes and Hands." *Knowledge and Society* 6:1–40.

———. 1988. "A Relativistic Account of Einstein's Relativity." *Social Studies of Science* 18:3–44.

———. 1993. *We Have Never Been Modern*. Trans. Catherine Porter. Cambridge, Mass.: Harvard University Press.

———. 1999. *Pandora's Hope*. Cambridge, Mass.: Harvard University Press.

Lawrence, D. H. 1964. *The Complete Poems*. Ed. Vivian de Sola Pinto and Warren Roberts. New York: Viking Press.

Leckman, James, and Donald Cohen. 1988. "Descriptive and Diagnostic Classification of Tic Disorders." In Cohen, Bruun, and Leckman 1988, 3–19.

Leckman, James, Mark Riddle, and Donald Cohen. 1988. "Pathobiology of Tourette's Syndrome." In Cohen, Bruun, and Leckman 1988, 103–16.

Lempers, J., E. Flavell, and J. Flavell. 1977. "The Development in Very Young Children of Tacit Knowledge concerning Visual Perception." *Genetic Psychology Monographs* 4.

Lethem, Jonathan. 1999. *Motherless Brooklyn*. New York: Doubleday.

Leung, E., and H. Rheingod. 1981. "Development of Pointing as a Social Gesture." *Developmental Psychology* 17:215–20.

Levinas, Emmanuel. 1989. "Ethics as First Philosophy." Trans. Seán Hand and Michael Temple. In *The Levinas Reader,* ed. Seán Hand, 75–87. Oxford: Blackwell.

———. 1996. "Meaning and Sense." Trans. Alphonso Lingis, rev. Simon Critchley and Adriaan Peperzak. In *Basic Philosophical Writings,* ed. Adriaan Peperzak, Simon Critchley, and Robert Bernasconi, 33–64. Bloomington: Indiana University Press.

Levine, George. 2006. *Darwin Loves You: Natural Selection and the Re-Enchantment of the World.* Princeton, N.J.: Princeton University Press.

Lévi-Strauss, Claude. 1966. *The Savage Mind,* no translator. Chicago: University of Chicago Press.

———. 1975. *The Raw and the Cooked.* Trans. John and Doreen Weightman. New York: Harper Torchbooks.

Lieberman, Philip. 1998. *Eve Spoke: Human Language and Human Evolution.* London: Picador.

———. 2000. *Human Language and Our Reptilian Brain: The Subcortical Bases of Speech, Syntax, and Thought.* Cambridge, Mass.: Harvard University Press.

MacLean, P. D., and J. D. Newman. 1988. "Role of Midline Frontolimbi Cortex in the Production of the Isolation Call of Squirrel Monkeys." *Brain Research* 450:111–23.

Martinet, André. 1962. *A Functional View of Language.* Oxford: Clarendon.

Matthiessen, F. O. 1941. *American Renaissance: Art and Expression in the Age of Emerson and Whitman.* New York: Oxford University Press.

Mayr, Ernst. 1982. *The Growth of Biological Thought.* Cambridge, Mass.: Harvard University Press.

McFadden, Johnjoe. 2000. *Quantum Evolution.* New York: Norton.

McManus, Chris. 2002. *Right Hand Left Hand: The Origins of Asymmetry in Brains, Bodies, Atoms, and Cultures.* Cambridge, Mass.: Harvard University Press.

McNeill, Daniel. 1998. *The Face.* Boston: Little Brown.

Melzack, Ronald. 1993. "Pain: Past, Present, Future." *Canadian Journal of Experimental Psychology* 47:615–29.

———. 1995. "Phantom-Limb and the Brain." In *Pain and the Brain: From Nociception to Cognition,* ed. Burkhart Bromm and John E. Desmedt. Advances in Pain Research and Therapy, vol. 22. New York: Raven Press.

Melzack, Ronald, and Patrick Wall. 1983. *The Challenge of Pain.* New York: Basic Books.

Middleton, F. A., and P. L. Strick. 1994. "Anatomical Evidence for Cerebellar and Basal Ganglia Involvement in Higher Cognition." *Science* 31:458–61.

Miller, James Q. 2001. "The Voice in Tourette Syndrome." *New Literary History* 32:519–36.

Miller, J. Hillis. 1985. "The Search for Grounds in Literary Study." In *Rhetoric and Form,* ed. Robert Con Davis and Ronald Schleifer, 19–36. Norman: University of Oklahoma Press.

———. 1986. *The Ethics of Reading.* New York: Columbia University Press.

Mithen, Steven. 2006. *The Singing Neanderthals: The Origins of Music, Language, Mind, and Body.* Cambridge, Mass.: Harvard University Press.

Moore, Walter. 1992. *Schrödinger: Life and Thought.* Cambridge: Cambridge University Press.

Morgan, Emma. 1995. *A Stillness Built of Motion: Living with Tourette's.* Florence, Mass.: Hummingwoman Press.

Morris, David. 1993. *The Culture of Pain.* Berkeley: University of California Press.

———. 1998. *Illness and Culture in the Postmodern Age.* Berkeley: University of California Press.

Morrison, Toni. 1988. *Beloved.* New York: Plume.

Murphy, C. M., and D. J. Messer. 1977. "Mothers, Infants, and Pointing: A Study of Gesture." In *Studies in Mother-Infant Interaction,* ed. Heinz Rudolph Schaffer, 323–54. New York: Academic Press.

Napier, John. 1991. *Hands.* Rev. Russell H. Tuttle. Princeton, N.J.: Princeton University Press.

Newberg, Andrew, and Eugene d'Aquili. 2001. *Why God Won't Go Away: Brain Science and the Biology of Belief.* New York: Ballantine Books.

Nietzsche, Friedrich. 1967a. *The Birth of Tragedy and The Case of Wagner.* Trans. Walter Kaufmann. New York: Vintage Books.

———. 1967b. *On the Genealogy of Morals and Ecce Homo.* Trans. Walter Kaufmann and R. J. Hollingdale. New York: Vintage Books.

Niiniluoto, Ilkka. 1999. "Defending Abduction." *Philosophy of Science* 66: S436–51.

O'Connor, Flannery. 1972. "The Artificial Nigger." In *The Complete Stories.* New York: Noonday Press.

Ornstein, Robert. 1997. *The Right Mind: Making Sense of the Hemispheres.* New York: Harcourt, Brace.

Pauls, D. L., J. F. Leckman, K. E. Towbin, G. E. P. Zahner, and D. J. Cohen. 1986. "A Possible Genetic Relationship Exists between Tourette's Syndrome and Obsessive-Compulsive Disorder." *Psychopharmacology Bulletin* 22:730–33.

Peirce, Charles Sanders. 1931–35. *Collected Papers.* Vols. 1–6. Edited by C. Hartshorne and P. Weiss. Cambridge, Mass.: Belknap Press, Harvard University.

———. 1992. "Deduction, Induction, and Hypothesis." In *The Essential Peirce,* ed. Nathan Houser and Kloesel, 186–99. Bloomington: Indiana University Press.

Pinker, Steven. 1994. *The Language Instinct.* New York: Harper Perennial.

Plotkin, Henry. 1994. *Darwin Machines and the Nature of Knowledge.* Cambridge, Mass.: Harvard University Press.

Poovey, Mary. 1998. *A History of the Modern Fact: Problems of*

Knowledge in the Sciences of Wealth and Society. Chicago: Chicago University Press.
Prigogine, Ilya, and Isabelle Stengers. 1984. *Order out of Chaos: Man's New Dialogue with Nature*. New York: Bantam Books.
Provine, Robert R. 2000. *Laughter: A Scientific Investigation*. New York: Penguin Books.
Ramachandran, V. S. 2004. *A Brief Tour of Human Consciousness*. New York: Pi Press.
Ramachandran, V. S., and Sandra Blakeslee. 1998. *Phantoms in the Brain: Probing the Mysteries of the Human Mind*. New York: Quill.
Rey, Roselyne. 1993. *The History of Pain*. Trans. Louise Elliott Wallace et al. Cambridge, Mass.: Harvard University Press.
Reynolds, Peter. 1993. "The Complementation Theory of Language and Tool Use." In Rita Gibson and Tim Ingold, *Tools, Language, and Cognition in Human Evolution*, 407–28. Cambridge: Cambridge University Press.
Ricoeur, Paul. 1984. *Time and Narrative*. Vol. 1. Trans. Kathleen McLaughlin and David Pellauer. Chicago: University of Chicago Press.
Robertson, Mary, and Simon Baron-Cohen. 1998. *Tourette Syndrome: The Facts*. Oxford: Oxford University Press.
Robinson, Richard. 1972. *Definition*. Oxford: Oxford University Press.
Rotman, Brian. 1987. *Signifying Nothing: The Semiotics of Zero*. Stanford, Calif.: Stanford University Press.
———. 1993. *Ad Infinitum, The Ghost in Turing's Machine: Taking God Out of Mathematics and Putting the Body Back In*. Stanford, Calif.: Stanford University Press.
Roubaud, Jacques. 1990. *Some Thing Black*. Trans. Rosemarie Waldrop. Elmwood Park, Ill.: Dalkey Archive Press.
Rubio, Gwyn Hyman. 1998. *Icy Sparks*. New York: Penguin Books.
Russell, Bertrand. 1917. *Mysticism and Logic*. Garden City, N.Y.: Doubleday.
———. 1923. *The ABC of Atoms*. London: Kegan Paul, Trench, Trubner.
———. 1954. *The Analysis of Matter*. New York: Dover.
Sacks, Oliver. 1987. *The Man Who Mistook His Wife for a Hat*. New York: Harper and Row.
———. 1989. *Seeing Voices: A Journey into the World of the Deaf*. Berkeley: University of California Press.
———. 1995. *An Anthropologist on Mars*. New York: Vintage Books.
———. 1999. *Awakenings*. New York: Vintage Books.
———. 2007. *Musicophilia: Tales of Music and the Brain*. New York: Knopf.
Said, Edward. 1994. "The Politics of Knowledge." In *Contemporary Literary Criticism*, ed. Robert Con Davis and Ronald Schleifer, 3rd ed., 144–53. New York: Longman.

Saussure, Ferdinand de. 1959. *Course in General Linguistics.* Trans. Wade Baskin. New York: McGraw Hill.

———. 1983. *Course in General Linguistics.* Trans. Roy Harris. La Salle, Ill.: Open Court Press.

Scarry, Elaine. 1985. *The Body in Pain: The Making and Unmaking of the World.* New York: Oxford University Press.

Schleifer, Ronald. 1987. *A. J. Greimas and the Nature of Meaning.* London: Croom Helm.

———. 1990. *Rhetoric and Death: The Language of Modernism and Postmodern Discourse Theory.* Urbana: University of Illinois Press.

———. 1993. "Rural Gothic: The Sublime Rhetoric of Flannery O'Connor." In *Frontier Gothic: Terror and Wonder at the Frontier in American Literature,* ed. David Mogen, Scott Sanders, and Joanne Karpinski, 175–86. Madison, N.J.: Fairleigh Dickinson University Press.

———. 1999. "'What Is This Thing Called Love?' Cole Porter and the Rhythms of Desire." *Criticism* 41:7–23.

———. 2000a. *Analogical Thinking: Post-Enlightenment Understanding in Language, Collaboration, and Interpretation.* Ann Arbor: University of Michigan Press.

———. 2000b. *Modernism and Time: The Logic of Abundance in Literature, Science, and Culture, 1880–1930.* Cambridge: Cambridge University Press.

Schleifer, Ronald, Robert Con Davis, and Nancy Mergler. 1992. *Culture and Cognition: The Boundaries of Literary and Scientific Inquiry.* Ithaca, N.Y.: Cornell University Press.

Schleifer, Ronald, and Nancy West. 1999. "The Poetry of What Lies Close at Hand: Photography, Commodities, and Post-Romantic Discourses in Hardy and Stevens." *Modern Language Quarterly* 60:33–57.

Schleifer, Ronald, and Jerry Vannatta. 2006. "The Logic of Diagnosis: Peirce, Literary Narrative, and the History of Present Illness." *Journal of Medicine and Philosophy* 31:363–85.

Scholes, Robert. 1985. *Textual Power: Literary Theory and the Teaching of English.* New Haven, Conn.: Yale University Press.

Schrödinger, Erwin. 1992. *What Is Life?* Cambridge: Cambridge University Press.

Searle, John. 1969. *Speech Acts: An Essay in the Philosophy of Language.* Cambridge: Cambridge University Press.

Shapin, Stephen. 1996. *The Scientific Revolution.* Chicago: University of Chicago Press.

Sheriff, John K. 1994. *Charles Peirce's Guess at the Riddle: Grounds for Human Significance.* Bloomington: University of Indiana Press.

Siegfried, Tom. 2000. *The Bit and the Pendulum: From Quantum Computing to M Theory—The New Physics of Information.* New York: Wiley.

Silverio de Santa Teresa, P. 1946. *The Complete Works of Saint Teresa of Jesus*. 3 vols. Trans. E. Allison Peers. London: Sheed and Ward.

Smith, David Livingstone. 2004. *Why We Lie: The Evolutionary Roots of Deception and the Unconscious Mind*. New York: St. Martin's Press.

Stafford, Jean. 1969. "The Interior Castle." In *The Collected Stories of Jean Stafford,* 179–93. New York: Farrar, Strauss, and Giroux.

Steiner, George. 1975. *After Babel*. New York: Oxford University Press.

Stevens, Janice R., and Paul H. Blachly. 1966. "Successful Treatment of the Maladie des Tics, Gilles de la Tourette's Syndrome." *American Journal of Diseases in Children* 112:541–45.

Stevens, Wallace. 1972. *The Palm at the End of the Mind: Selected Poems and a Play*. Ed. Holly Stevens. New York: Vintage.

Sutton, D., and U. Jurgens. 1988. "Neural Control of Vocalization." In *Comparative Primate Biology,* ed. H. D. Stiklis and J. Erwin, 4:635–47. New York: Arthur D. Liss.

Towbin, Kenneth. 1988. "Obsessive-Compulsive Symptoms in Tourette's Syndrome." In Cohen, Bruun, and Leckman 1988, 137–49.

Trimble, Michael R. 2007. *The Soul in the Brain: The Cerebral Basis of Language, Art, and Belief*. Baltimore: Johns Hopkins University Press.

Vannatta, Jerry, Ronald Schleifer, and Sheila Crow. 2005. *Medicine and Humanistic Understanding: The Significance of Narrative in Medical Practices*. Philadelphia: University of Pennsylvania Press.

Veblen, Thorstein. 1932. *The Theory of Business Enterprise*. New York: Scribner.

Vertosick, Frank. 2000. *Why We Hurt: The Natural History of Pain*. New York: Harcourt.

Wall, Patrick. 1999. *Pain: The Science of Suffering*. London: Phoenix Books.

Warminski, Andrzej. 1987. *Readings in Interpretation*. Minneapolis: University of Minnesota Press.

———. 1996. "Introduction: Allegories of Reference." In *Paul de Man, Aesthetic Ideology,* ed. Andrzej Warminski, 1–33. Minneapolis: University of Minnesota Press.

Weil, Simone. 2001. *Waiting for God*. Trans. Emma Craufurd. New York: Perennial Classics.

Weinberg, Steven. 1992. *Dreams of a Final Theory: The Search for the Fundamental Laws of Nature*. New York: Pantheon Books.

Weiner, Jonathan. 1995. *The Beak of the Finch*. New York: Vintage.

Wheeler, John. 1986. Interview. In Davies and Brown 1986a, 55–69.

Wiener, Norbert. 1961. *Cybernetics*. Cambridge: MIT Press.

———. 1967. *The Human Use of Human Beings*. New York: Avon Books.

Williams, Raymond. 1961. *The Long Revolution*. Harmondsworth, England: Penguin.

Wilson, David Sloan. 2003. *Darwin's Cathedral: Evolution, Religion, and the Nature of Society.* Chicago: University of Chicago Press.

Wilson, E. O. 1975. *Sociobiology: The New Synthesis.* Cambridge, Mass.: Harvard University Press.

———. 1998. *Consilience: The Unity of Knowledge.* New York: Knopf.

Wilson, Frank R. 1999. *The Hand: How Its Use Shapes the Brain, Language, and Human Culture.* New York: Vintage Press.

Wittgenstein, Ludwig. 2006. *Philosophical Investigations,* 3rd ed. Trans. G. E. M. Anscombe. Oxford: Blackwell.

Woodger, J. H. 1930. "The 'Concept of Organism' and the Relation between Embryology and Genetics." *Quarterly Review of Biology* 5:1–22.

Zuckerkandl, Victor. 1969. *Sound and Symbol: Music and the External World.* Trans. Willard Trask. Princeton, N.J.: Princeton University Press.

Index

Ronald Schleifer is George Lynn Cross Research Professor of English and adjunct professor in the College of Medicine at the University of Oklahoma. His other books include *Modernism and Time: The Logic of Abundance in Literature, Science, and Culture, 1880–1930* and *Analogical Thinking: Post-Enlightenment Understanding of Language, Collaboration, and Interpretation.*